Short Studies from the Walter Chapin Simpson Center for the Humanities
The Center for the Humanities at the University of Washington was established in 1987 with a mandate to encourage interdisciplinary activities in the humanities. Supported by the University Initiative Fund and endowed in 1997 in the name of Walter Chapin Simpson, the Center is dedicated to fostering innovative research and teaching in the humanities, and to stimulating exchange and debate on cultural and educational issues, both on and off the University of Washington campus. Its broader goal is to knit together the academic and civic communities through a shared fostering of education and culture. The Simpson Center sponsors a wide range of activities, including interdisciplinary courses and collaborative research groups, public lectures, symposia, arts events, publications, and a fellowship program for University of Washington faculty and graduate students.

Semiotic Flesh: Information & the Human Body

is a volume in the new series Short Studies

from the Walter Chapin Simpson Center for the Humanities.

Semiotic Flesh:

Information & the Human Body

Edited by Phillip Thurtle and Robert Mitchell

Copyright © 2002 by the Walter Chapin Simpson Center for the Humanities

Published by the Walter Chapin Simpson Center for the Humanities
University of Washington
Seattle, USA

Library of Congress Cataloging-in-Publication Data
Semiotic Flesh: Information and the Human Body / edited by Phillip Thurtle and Robert Mitchell.
          p. cm. — (Short Studies from the Walter Chapin Simpson Center for the Humanities)
Includes bibliographical references.
ISBN 0-295-98200-4
1. Body, human. 2. Computer science—Social aspects. 3. Information technology—Social aspects. I. Thurtle, Phillip. II. Mitchell, Robert (Robert Edward), 1969- III. Series.

HM636 .S46 2002
306.4—dc21

2001054208

Series editors
Kathleen Woodward
Margit A. E. Dementi

Series Design
Christopher Ozubko

Printed in the United States of America
Distributed by the University of Washington Press
PO Box 50096
Seattle, Washington 98145-5096
USA

The paper used in this publication meets the minimum requirements of American National Standard for Information Sciences—Permanence of Paper for Printed Library Materials, ANSI Z39.48-1984.

Contents

ACKNOWLEDGMENTS

In addition to the contributors and respondents for this volume, we would also like to thank the Walter Chapin Simpson Center for the Humanities, for its financial support of our lecture series and this volume (and an especially heartfelt thanks to Kathleen Woodward, Margit Dementi, and Liz Browning for their unflagging support and optimism); the Comparative History of Ideas (CHID) program for its support of Robert Mitchell's undergraduate course, "Semiotic Flesh: A Genealogy of the Concept of Information"; the University of Washington departments of Communications, Comparative Literature, and History, and the Elizabeth S. Soule Endowed Professorship in Nursing for their financial support of our lecture series; Jacquie Ettinger, for her patience and help; and Tyler Fox, for his help, which went above and beyond the call of duty, in providing logistical support for the lecture series.

Introduction

Fleshy Data: Semiotics, Information, and the Body

Robert Mitchell and Phillip Thurtle

It seems so easy to keep separate: on one side is information and on the other is flesh and blood. Information is that which exists between elements; bodies are the elements themselves. One is abstract, the other corporeal. One is intricately involved in signs and syntax, the other in cells and organs. One operates through the metaphysics of absence, the other through the metaphysics of presence. For much of the twentieth century, this apparently solid conceptual wall between bodies and information has allowed us to separate these two aspects of modern life.

Yet in the last few decades, it has become increasingly clear that this conceptual wall leaks—bodies and information will not stay separate from one another. Moreover, one begins to suspect that the distinction between bodies and information has obscured as many important questions as it purported to solve. This distinction makes it difficult to understand, for example, why one of the largest government-sponsored biomedical research agendas is concerned with an informatics of the human body.[1] And why are numerous bioethicists puzzling about social or personal ownership of genetic information? What is the "view of life" that allows some molecular biologists to claim that molecular biology is an information science and has been from the very beginning? Why do many popular entertainment forms imagine an extra-bodily existence in a "matrix" of information? Why do computer and video games allow us to selectively augment body parts, skills, and powers with the sleight of hand of digital representation? Why are some of the most provocative artists using biological materials in artistic expression?

Why, in short, do information and human bodies refuse to stay on their respective sides of the conceptual wall? And what can we learn from this?

This book demonstrates that the theoretical osmosis between bodies and information in fact indicates new ways of inhabiting, imagining, and manipulating bodies. These new forms of embodiment call for a new imaginary, or better yet, virtual landscape to help us navigate the future of our becoming.[2] Emblematic of this attempt to reimagine the relationships between bodies and information is the suggestion by Case, the protagonist of William Gibson's classic cyberpunk novel *Neuromancer*, that advanced information technologies enable the possibility of "data made flesh."[3] This provocative phrase condenses a number of hopes and fears associated with the relationship between information and the flesh. One can read it as indicative of a troubling trend in the sciences, in which flesh and bodies are losing out to abstract notions of information.[4] This interpretation highlights justified fears that we spend more time engaged with experiences generated by informational technologies than in personal interactions. Or the phrase might suggest that we are entering a period in which we have the capabilities to use informational technologies to generate new forms of fleshy experience.[5] This interpretation points to the powerful use of informational technologies in redefining current experiences of our embodied self and others. Recent developments in the cloning of mammals are only the most literal example of these new experiences.

Although much in both of these interpretations is correct, they do not provide the complete story. Missing is an examination of that critical instant when information becomes flesh, a moment we can think of as the material poiesis of informatics. Much depends on understanding this moment. In what specific ways do we enact the transformation from the virtual world of information to the actual world of flesh and bone, and vice versa? What are the implications of this moment for human biological existence (the species body), communal interactions (the social body), or the accessing and transformation of shared memory (the body of knowledge)?

The three essays included in this volume all focus on different aspects of this informatic poiesis, that moment when flesh and information merge and begin to signify. Each essay was originally delivered as part of a lecture series entitled "Information and the Human Body," held at the University of Washington in January and February of 2001. The series was funded in largest part by the Walter Chapin Simpson Center for the Humanities, which has also underwritten this publication.[6] This volume is designed to be an enduring record of the talks and the exciting discussions that followed them. In the spirit of providing a record of these exchanges, we have commissioned a short response to each essay from researchers at the University of Washington.

In addition to the financial and logistical support provided by an interdisciplinary mixture of groups at the University of Washington, the essays and responses in this volume draw on an important intellectual lineage that collectively suggests an emerging area of study. While this field has yet to receive a proper name, we tentatively might call it "information studies." Early researchers in this field include historians such as Mark Poster, who in the late 1970s introduced the notion of the "mode of information" as a tool for historical study,[7] and sociologists such as Manuel Castells, who has sought to describe the dynamics of the Informational City and the Network Society.[8] Each of the three primary authors included in this volume has also helped demarcate the field of information studies. The important work of understanding the relationship of informational concepts to embodiment as developed through the field of molecular biology has been advanced by Richard Doyle and Timothy Lenoir,[9] and N. Katherine Hayles has sought to untangle the very complicated discourses that link biology, cybernetics, chaos theory, and the fiction of authors such as William S. Burroughs and Thomas Pynchon.[10] The recent publication of a number of accounts of information intended for larger lay audiences (for example, Michael E. Hobart and Zachary S. Schiffman's *Information Ages: Literacy, Numeracy, and the Computer Revolution*) suggests that this field may be gaining more widespread recognition.[11]

The three primary essays in this volume constitute the most recent contributions by Doyle, Hayles, and Lenoir to the important task of understanding the relationship between bodies, information, and meaning. In "LSDNA: Consciousness Expansion and the Emergence of Biotechnology," Richard Doyle continues work that he began in *On Beyond Living*. In that text, Doyle outlined the implications of an informationalized understanding of DNA for the concept of "life." He suggested there that the notion of "life," which seemed to require a vitalist belief in its own unplumbable depths, might fall victim to the transparency assumed by the term "information." But he also hinted at ways in which the notion of life was itself transforming in the wake of information and information technologies (for example, via the notion of artificial life), and he has continued that exploration in his forthcoming *Wetwares! Experiments in Post-Vital Living*, as well as in his essay here. In "LSDNA," Doyle considers the forms of agency that have characterized several key relationships between scientific experimentation, information, and life in the past century. In his discussion of the strange connection between Albert Hofmann (inventor of LSD) and Kary Mullis (inventor of PCR, a DNA duplication technology), Doyle highlights paradoxical forms of agency (laughter, terror, ecstasy) that put into question notions of self-control and individuality so central to the popular image of the biological scientist. He also points to a very important change in the notion of information in biological discourse, as it has moved from a cryptographic paradigm (in which information was understood as the space of a revelation) to a pragmatic paradigm (in which one manipulates information to see what it can do, with little or no concern with what it might mean).

In "The Virtual Surgeon: Operating on the Data in an Age of Medialization," historian of biology Timothy Lenoir provides us with an example of his most recent research interest. Lenoir began his career with *The Strategy of Life: Teleology and Mechanics in Nineteenth Century German Biology*, in which he considered concepts of evolution, before turning to an analysis of more contemporary institutionalizations of scientific knowledge in *Instituting Science: The Cultural Production of Scientific Disciplines*.[12] More recently, he has investigated the intersection of computing and biomedical discourses, concentrating especially on the application of computing to surgery, and in the essay for this volume, he considers the ways in which recent developments in surgery have enabled a "fusion of digital and physical reality" (28), rather than the abstraction from the flesh that some critics have warned against. He notes that "medialized" surgery is in fact far less abstract than surgery accomplished with pre-informatic tools (so, for example, craniofacial surgery formerly required the use of "ideal" skulls to which surgeons hoped patients' actual physiology conformed; highly medialized craniofacial surgery, by contrast, uses computers to generate a new personalized model for each patient). Lenoir notes that the cost of this fleshy data is increasing surveillance, for these informatic tools have been developed in part to facilitate the goals of "a massive system of preventive health care from genome to lifestyle" (41).

N. Katherine Hayles's "Flesh and Metal: Reconfiguring the Mindbody in Virtual Environments" represents the latest development in a sustained reflection on art and science that dates back to Hayles's first book, *The Cosmic Web: Scientific Field Models and Literary Strategies in the Twentieth Century*. The essay emerges more directly from Hayles's recent and extremely influential text *How We Became Posthuman: Virtual Bodies in Cybernetics, Literature, and Informatics*, in which she sought to show how information became disembodied in the 1950s, as well as what it might mean to become "posthuman." In "Flesh and Metal," Hayles picks up where *How We Became Posthuman* left off, as she investigates the role of the human body in three recent virtual reality (VR) installations. As she notes in her first section, the essay emerged in part as a response to criticisms that the distinction between "the body" and "embodiment," articulated in *How We Became Posthuman*, still depended upon a Cartesian dualism. In order to counter this reading, Hayles develops Mark Hansen's notion of the mindbody in order to consider the following VR installations: *Traces*, by Simon Penny and collaborators; *Einstein's Brain*, by Alan Dunning, Paul Woodrow, and collaborators; and *NØtime*, by Victoria Vesna and her collaborators. Hayles argues that all three works emphasize the non-Cartesian mindbody, but each emphasizes a different aspect: *Traces* investigates "the relation of mindbody to the immediate surroundings";

Einstein's Brain highlights perception; and *NØtime* "emphasizes relationality as cultural construction" (53).

Our three respondents have also played an important role in helping to demarcate the emerging field of information studies. In his pathbreaking text *The Cinematic Body*, Steven Shaviro (professor of English at the University of Washington) highlighted the role of the body and affect in the experience of film viewing, and in his more recent volume *Doom Patrols: A Theoretical Fiction about Postmodernism*, he extends that approach to other media.[13] In his response to "LSDNA," Shaviro attempts to draw lessons from the unexpected connections between LSD and DNA that Doyle locates, concluding that the links between "[g]ene-splicing and consciousness-altering activities compel us to concede more to the biochemical realm than we might otherwise want to (since they wreak havoc upon our ideas of free will and responsibility) and yet to reject any sort of biological determinism (since they teach us how to induce biochemical changes at will—even if we are not in control of the consequences)" (27).

Our second respondent, Peter Oppenheimer, has had a very practical impact on the realm of information studies; a VR researcher at the Human Interface Technology (HIT) Lab at the University of Washington, Oppenheimer currently works on some of the surgical applications of VR described in Lenoir's essay. Oppenheimer takes up Lenoir's notion of a blending of the virtual and the real, but through the example of computer-driven "replicators," notes that the union between the real and virtual may extend even further than Lenoir himself suggests. Oppenheimer also urges us to locate virtual perspectives on apparently real objects, while being attentive to the emergence of new forms and entities that confound the dualism of the virtual and the real.

Finally, Kathleen Woodward, Director of the Walter Chapin Simpson Center for the Humanities at the University of Washington, has written and edited several important books on aging and obsolescence,[14] as well as several texts on the intersections between literature and information technologies.[15] More recently, Woodward has sought to understand the connection between information technologies and the emotions,[16] and in her response to Hayles's essay she continues that effort. She finds useful Hayles's adoption of the notion of the "lifeworld," as well as Hayles's stress on "interaction, mutuality, and codetermination," but she argues that we should also try to recognize that emotions "serve as a prosthesis connecting the technological world of virtual reality with the humanworld" (70).

Each of the essays and responses in this volume highlights the fact that we have entered an era when signifying practice and embodiment are no longer conceptually or practically separate. When we can utilize informational technologies to directly manipulate biological makeup, when living agents are considered another medium for creative expression, and when the dissemination of biomolecular choices has made older moral guidance obsolete: the membrane-thin distinction between semiotics and flesh has failed us. We thus offer this volume as collected memories of the ways that "information" emerged in the West, a demonstration of how it has changed our concept of embodiment, and a map of how widely this concept resonates in different fields.

---

1   In fiscal year 1999, for instance, the federal appropriations for the Human Genome Project totaled $315.5 million. See Human Genome Program, U.S. Department of Energy, *Human Genome News* 10 (1999): 3–4.

2   We use the term "virtual" in the sense outlined in Gilles Deleuze, *Difference and Repetition*, trans. Paul Patton (New York: Columbia University Press, 1995), pp. 191–214.

3   William Gibson, *Neuromancer* (New York: Ace Books, 1984), p. 16.

4   In her recent history of the posthuman, N. Katherine Hayles reads Gibson's quote in this manner, suggesting that it emphasizes "cognition rather than embodiment." See *How We Became Posthuman: Virtual Bodies in Cybernetics, Literature, and Informatics* (Chicago: University of Chicago Press, 1999), p. 5.

5   Richard Doyle reads Gibson's quote in this manner in *Wetwares! Experiments in Post-Vital Living* (Minneapolis: University of Minnesota Press, forthcoming).

6   We are pleased that this is the first volume of the Simpson Center's Short Studies to represent a lecture series held at the University of Washington.

7   Mark Poster, *Foucault, Marxism, and History: Mode of Production versus Mode of Information* (Cambridge: Polity Press, 1984). See also Poster's *The Mode of Information: Poststructuralism and Social Context* (Chicago: University of Chicago Press, 1990) and *The Second Media Age* (Cambridge: Polity Press, 1995).

8   See Manuel Castells, *The Informational City: Information Technology, Economic Restructuring, and the Urban-Regional Process* (New York: Basil Blackwell, 1989) and *The Information Age: Economy, Society, and Culture*, 3 vols. (New York: Basil Blackwell, 1996–98). More distant, but equally enriching, predecessors in the field of information studies include Hans Ulrich Gumbrecht (through his emphasis on the "materialities of communication"); Joshua Meyrowitz, Marshall McLuhan, and Harold Innis (for their explication of medium theory); Eric Havelock, Walter Ong, Elizabeth Eisenstein, Jack Goody, and Adrian Johns (for their discussions of the dynamics of orality and literacy); Elizabeth Grosz (for her recognition of the importance of "embodiment" and "becoming" for feminism); Eric Michaels (through his evocative charting of informational flows in Australian Aboriginal communities); Friedrich Kittler (through a provocative mix of discourse analysis, McLuhanian medium theory, and Lacanian theories of identification);

Erving Goffman (especially his argument for identity management as offered in *Stigma*); Kathleen Woodward (through her critique of information and important emphasis on the aging body in technologically centered societies); Gregory Bateson, Humberto Maturana, and Francisco Varela (for their discussion of autopoiesis); Alphonso Lingis (through his rethinking of the phenomenological tradition in the direction of embodiment); and Gilles Deleuze, especially the important works *Difference and Repetition* and *The Logic of Sense*, trans. Mark Lester with Charles Stivale (New York: Columbia University Press, 1990).

9   See Richard Doyle, *On Beyond Living: Rhetorical Transformations of the Life Sciences* (Stanford: Stanford University Press, 1997); Timothy Lenoir, "Shaping Biomedicine as an Information Science," in *Proceedings of the 1998 Conference on the History and Heritage of Science Information Systems*, ed. Mary Ellen Bowden, Trudi Bellardo Hahn, and Robert V. Williams, ASIS Monograph Series (Medford, New Jersey: Information Today, Inc., 1999), pp. 27–45, and "Visions of Theory: Fashioning Molecular Biology as an Information Science," to appear in M. Norton Wise, *Growing Explanations* (Stanford: Stanford University Press, forthcoming). Also of interest in this connection is the work of Lily E. Kay, especially *The Molecular Vision of Life: Caltech, the Rockefeller Foundation, and the Rise of the New Biology* (Oxford: Oxford University Press, 1993) and *Who Wrote the Book of Life? A History of the Genetic Code* (Stanford: Stanford University Press, 2000).

10 In addition to *How We Became Posthuman*, see *The Cosmic Web: Scientific Field Models and Literary Strategies in the Twentieth Century* (Ithaca: Cornell University Press, 1984) and *Chaos Bound: Orderly Disorder in Contemporary Literature and Science* (Ithaca: Cornell University Press, 1990).

11 Michael E. Hobart and Zachary S. Schiffman, *Information Ages: Literacy, Numeracy, and the Computer Revolution* (Baltimore: Johns Hopkins University Press, 1998). Also deserving of notice are the following: Albert Borgmann, *Holding on to Reality: The Nature of Information at the Turn of the Millennium* (Chicago: University of Chicago Press, 1999); James J. O'Donnell, *Avatars of the Word: From Papyrus to Cyberspace* (Cambridge: Harvard University Press, 1998); Richard A. Lanham, *The Electronic Word: Democracy, Technology, and the Arts* (Chicago: University of Chicago Press, 1993); Michael Heim, *Electric Language: A Philosophical Study of Word Processing* (New Haven: Yale University Press, 1987) and *Virtual Realism* (New York: Oxford University Press, 1998); Richard Coyne, *Designing Information Technology in the Postmodern Age: From Method to Metaphor* (Cambridge: MIT Press, 1995) and *Technoromanticism: Digital Narrative, Holism, and the Romance of the Real* (Cambridge: MIT Press, 1999).

12 Timothy Lenoir, *The Strategy of Life: Teleology and Mechanics in Nineteenth Century German Biology* (Boston: D. Reidel, 1982) and *Instituting Science: The Cultural Production of Scientific Disciplines* (Stanford: Stanford University Press, 1997).

13 Steven Shaviro, *The Cinematic Body* (Minneapolis: University of Minnesota Press, 1993) and *Doom Patrols: A Theoretical Fiction about Postmodernism* (New York: Serpent's Tail, 1997).

14 See Kathleen Woodward, *Aging and Its Discontents: Freud and Other Fictions* (Bloomington: Indiana University Press, 1991), and Woodward, ed., *Figuring Age: Women, Bodies, Generations* (Bloomington: Indiana University Press, 1999).

15 See Kathleen Woodward, ed., *The Myths of Information: Technology and Postindustrial Culture* (Madison, Wisconsin: Coda Press, 1980), and Teresa de Lauretis, Andreas Huyssen, and Kathleen Woodward, eds., *The Technological Imagination: Theories and Fiction* (Madison, Wisconsin: Coda Press, 1980).

16 See, for example, Kathleen Woodward, "Statistical Panic," *differences: A Journal of Feminist Cultural Studies* 11, no. 2 (1999): 177–203.

LSDNA: Consciousness Expansion and the Emergence of Biotechnology

Richard Doyle

I had to struggle to speak intelligibly.
(Albert Hofmann on his self-experiment with LSD-25)

Finding a place to start is of utmost importance. Natural DNA is a tractless coil, like an unwound and tangled audio tape on the floor of the car in the dark.
(Kary Mullis on the invention of polymerase chain reaction)

In the schematic argument that follows, I want to map the rather willy-nilly itinerary of molecular biology's rhetorical and conceptual evolution and its debts to those forms of agency best exemplified by laughter but available to many extraordinary affects "proper" to even the scientific will: laughter, terror, ecstasy. Such modes of response, I will argue, were crucial to a conceptual evolution whose feedback loop arrives at cloning and tends toward a nanotechnological impasse: DNA information, at first understood by molecular biology as a fundamentally stable semantic phenomenon or "secret," becomes a spectacularly mutable technology of replication and differentiation by the early 1980s. This undoing of life, and the concomitant "loss" of integrity in the organism—wrought first by recombinant DNA and then by polymerase chain reaction—seems to occur in response to an undoing and doing of identity sculpted by that most tabooed and double-entendred scientific enterprise: the self-experiment. In particular, the necessary role of the self-experiment in the scientific study of hallucinogens—an inquiry not into life but into consciousness—provides the ecology for the emergence of these innovative and even ecstatic modes of interaction.

My discussion will respond to the common self-experiments at play in the seemingly diverse ecologies of the hallucinogenic "expansion" of "consciousness" and the engineering of evolution, biotechnology. These ecologies will be treated as twin or replicated domains where an informatic desire distributes and disperses both consciousness and life into inhuman, inorganic, and extraterrestrial realms.

Taking DNA, or Tripping Over the Organism

Thus far I have claimed a shift from an emphasis on the interiority of organisms and the associated revelation of their essence to a harnessing of DNA's capacity to replicate and its subsequent "distribution" of life. No longer simply the attribute of a sovereign organism, life now emerges out of the connections of a network, involving an essential impropriety—it is life's habit of refusing containment that becomes interesting for biotechnology and capital.

These replicants are extraordinarily different models of living systems, both of which take place under the sign of "DNA" and molecular biology. One, the cracking of the code, looked to expose vitality as an attribute of "a periodic crystal," an orderly rather than mysterious enterprise best apprehended by physics. Another, a model of living systems we associate with biotechnology and its collateral capital and publicity markets, is less interested in what DNA might "mean" than with what it can do, and the relation between what it can do—replicate—and its production of value. Indeed, an emphasis on the primordial importance of copying reminds us of the impossibility of keeping secrets, as replication allows alleles and ideas to travel to multiple contexts, some intended and some not. Contemporary life science, despite all the chatter about God associated with the recent rough draft of the human genome, is interested less in predictability than in experimentation and mutability, the capacity for deterritorialization that generates value in the economy.[1]

While molecular biology was busy on its eighth day of creation, discovering, decoding, and analyzing the "secret" of life, LSD-25 was also proffered in the labs and then the communes and other crowds of the world as the secret of "consciousness." Contrary to its usual representation as a seamless technology of unveiling associated with the instant gratification allegedly sought by an entire decade—the 1960s—LSD was continually treated as an enormously powerful but equally unreliable tool for the probing, revelation, and "expansion" of consciousness. Writers, researchers, and experimentalists such as Timothy Leary, Humphrey Osmond, Richard Alpert, and Stanislov Grof all sought to study the function of "set and setting" in the instantiation of hallucinogenic practices and their capacity to transform human consciousness. These writers took DNA and used it to frame and articulate hallucinogenic sessions as programmable but not controllable events, experiments with and on the self. As such, theirs was a fundamentally pragmatic rather than semantic relation to the "information" of DNA, more recipe than message.

How did these writers take DNA? In fact, there were many experiments among researchers attempting to ingest DNA and RNA itself as a hallucinogen, sometimes in the hope of developing a "learning lozenge" which would inscribe the experience of LSD onto the brain. But nucleic acids were also crucial rhetorical vectors composing hallucinogenic discourse of the 1950s and 1960s: the talk, thought experiments, manuals, and technical papers that resulted from variously intentional and unintentional ingestions.

Hallucinogenic discourse, both scientific and "recreational," faced a rhetorical impasse, one shared with the rest of the ecstatic tradition to which it responds: it must report on an event which is in principle impossible to communicate. Writers of mystic experience from St. Theresa to William James have treated the unrepresentable character of mystic events to be the very hallmark of ecstasy. Hallucinogenic discourse faced a similar struggle in the effort to report on the knowledge beyond what Aldous Huxley (and Jim Morrison) described as the "doors of perception." To deal with this struggle, many researchers had recourse to the rhetoric of nucleic acids—DNA and RNA became privileged characters in the stories and practices of hallucinogenic science. Nucleic acids were more, though, than "content providers" for the channeling of hallucinogenic knowledges into quasi-scientific protocols. As carriers of the news of molecular biology's informatic vision, rhetorics of nucleic acids were also set and setting for hallucinogenic sessions themselves: more than reporting devices, these rhetorics of nucleic acids helped to suggest that the sessions were, like DNA, programmable.

Double Take

Problems of reportage troubled the discourse of LSD almost from its very inception, and certainly from its very first ingestion. In his fundamental ergot studies in 1938, Albert Hofmann first synthesized lysergic acid diethylamide, abbreviated LSD-25 (Lysergsure-Dithylamid) for laboratory usage. This novel molecule was primarily noted for its strong effects on the uterus, but as Hofmann retroactively reports in his autobiography named for a molecule, *LSD: My Problem Child*, "the research report also noted, in passing, that the experimental animals became restless during the narcosis. The new substance, however, aroused no special interest in our pharmacologists and physicians; testing was therefore discontinued."[2] Hofmann continued his work in the ergot field, but LSD-25 was thought to have little pharmacological value, so between 1938 and 1943 "nothing more was heard of the substance LSD-25."

And yet Hofmann still had ears for the crying of LSD-25. According to his own account, LSD wouldn't leave him alone:

> And yet I could not forget the relatively uninteresting LSD-25. A peculiar presentiment—the feeling that this substance could possess properties other than those established in the first investigations—induced me, five years after the first synthesis, to produce LSD-25 once again so that a sample could be given to the pharmacological department for further tests. This was quite unusual; experimental substances, as a rule, were definitely stricken from the research program if once found to be lacking in pharmacological interest. . . . Nevertheless, in the spring of 1943, I repeated the synthesis of LSD-25. As in the first synthesis, this involved the production of only a few centigrams of the compound. (14)

Although Hofmann's attribution of presentiment must be placed within its context as an autobiographical confession, it nonetheless well names a peculiar agency that often adheres to those self-experiments that are survived: the inability to forget. The memory of LSD and, as we shall see, LSD-25 itself, seems to have little truck with the usual operations of will. Indeed, according to Hofmann, his response to the crying of LSD-25—synthesis—resulted paradoxically in an interruption or a dissolution: "In the final step of the synthesis, during the purification and crystallization of lysergic acid diethylamide in the form of a tartrate (tartaric acid salt), I was interrupted in my work by unusual sensations"(15). This interruption of the I, rather than ending an experiment, instigates one. In a report to a superior, Hofmann did his best to offer a description of the phenomenon that ensued after the interruption, but he only "surmised" a connection with the LSD-25.

Hofmann's analysis of the cause of the interruption was itself interrupted by the question of ingestion. Given his method of synthesis, Hofmann reasoned that the accidental passage must have been through his skin, and that the substance must therefore be of extraordinary—indeed, unprecedented—potency. Thus Hofmann's causal analysis of the unusual sensations seemed to offer two extraordinarily unlikely—i.e., unprecedented—alternatives. On the one hand, the cause could remain unknown, and the tasteless and odorless LSD-25 had merely been associated with the experience rather than causing it. In this instance the strange interruption retained its enigmatic status, a non sequitur of Hofmann's experience. On the other hand, if the minute quantity of LSD-25 were the causal agent of the interruption, then Hofmann was faced with the equally unlikely scenario that he had discovered the most potent compound known to history. To resolve the situation, Hofmann had recourse to another extraordinary non sequitur, itself seemingly emerging without cause: "There seemed to be only one way of getting to the bottom of this. I decided on a self-experiment" (16).

In what sense can one "decide on a self-experiment"? What warrants this decision? If it seems obvious that indeed, Hofmann did decide on such a course of action, it must also be noted that this decision is of necessity itself an experiment, one that emerges not from any deliberative logic but from an incalculable action, a break-age in a chain of reasoning: just do it. Hofmann's deliberation on his possible responses to the interruption was itself not subject to anything like a procedure, an algorithm shorter than repetition by which he could arrive at a resolution of the two equally enigmatic if thoroughly differentiated outcomes.[3] Thus the decision to self-experiment is itself a testing of a hypothesis: that repetition is the most high-fidelity and compressed procedure for resolving the matter. The outcome of this experiment— Hofmann's implication of himself into the research—cannot be meaningfully differ-entiated from the experimental dosing of LSD-25. Only an additional experiment— the synthesis and ingestion of LSD-25—will retroactively provide this experiment in decision making with anything like a result. As if to mark the extreme danger that the implication of a self and body into the experiment entails, Hofmann writes oxymoronically of a self-experiment embarked upon with "caution":

> Exercising extreme caution, I began the planned series of experiments with the
> smallest quantity that could be expected to produce some effect, considering
> the activity of the ergot alkaloids known at the time: namely, 0.25 mg (mg =
> milligram = one thousandth of a gram) of lysergic acid diethylamide tartrate. (16)

The danger, here, though, is not only the risk of an unknown compound of apparently extraordinary psychic potency. Rather, the instance or event of danger emerges coincident to the interruption of work itself: how to proceed? For as Hofmann notes after the now deliberate ingestion, LSD-25 is nothing if not the incessant and yet irregular arrival of the question, how to go on? This is a question of endurance for the experimental self: usual experimental protocol demands that everything is involved in an experiment except the self, and yet here it is precisely only the self and its responses that are the very assay of LSD-25. Only the variable examples of history provided anything like a protocol for self-experiment, calibration for the assay.[4] This assay had great difficulty generating any readout:

> 4/19/43 16:20: 0.5 cc of 1/2 promil aqueous solution of diethylamide tartrate orally
> = 0.25 mg tartrate. Taken diluted with about 10 cc water.
> Tasteless.
> 17:00: Beginning dizziness, feeling of anxiety, visual distortions, symptoms of
> paralysis, desire to laugh. (16)

LSD-25 did little to present itself for inscription. Without flavor, within forty minutes it produces predominantly anticipatory symptoms, events which were about to make themselves more fully known. The "desire to laugh," as such an experimental anticipatory symptom, was particularly difficult to assay. For by what means would any desire to laugh be registered by an observing self, except by laughter itself and its subsequent attempted blockage? What agency would interrupt said desire and, interrupted, in what sense did one desire to laugh?

Obviously, anyone can wish for laughter: this is the unlikely hope marked by the sitcom laugh track. But what is named by Hofmann is perhaps less a wish than a tantalizing inclination, a becoming laughter that is neither a cackling nor its absence, a meanwhile in which the proximity of the future—I am about to laugh—is unbearable to the self of the present. For Hofmann, this desire indeed becomes unbearable, the doing of the experiment veritably undone:

> Supplement of 4/21: Home by bicycle. From 18:00–ca. 20:00 most severe crisis. (See special report.)
> Here the notes in my laboratory journal cease. I was able to write the last words only with great effort. By now it was already clear to me that LSD had been the cause of the remarkable experience of the previous Friday, for the altered perceptions were of the same type as before, only much more intense. I had to struggle to speak intelligibly. (16)

Hofmann's certainty regarding the causal role of LSD-25 in the visions and disturbances he experienced was equaled by his inability to communicate the character and nature of the struggle. While under the variable influence of LSD-25, Hofmann periodically ceases to be capable of even an attempt at communication and thus, an attempt at experimentation. In place of the usual and invisible expectation of communicability that is the rhetorical arena of scientific observation, the struggle suggests Hofmann's reliance on another mode of knowing altogether: the struggle of the ordeal, an event in which knowing is not separable from an irreducible participation. "I had to struggle to speak intelligibly."

This inability to communicate is not a deficit in the hallucinogenic experience, but a symptom of it. In this sense, the self-experiment is both failure and success: as an experiment with the self, the outcome is close to null. No meaningful report can be generated, and therefore the knowledge of the hallucinogenic experience can in no way be gathered or repeated. As an assay, the self is found wanting. If the experiment is an occasion at which, strangely, the self is to be present as the very apparatus through which the inquiry is made visible and replicable, then the apparatus has faltered and the experiment is nothing but artifact.

And yet later, Hofmann would nonetheless prepare a special report to his supervisor, Professor Stoll. Here Hofmann was struck by the capacity to remember the experience with great, even machinic precision:

> What seemed even more significant was that I could remember the experience of LSD inebriation in every detail. This could only mean that the conscious recording function was not interrupted, even in the climax of the LSD experience, despite the profound breakdown of the normal world view. For the entire duration of the experiment, I had even been aware of participating in an experiment, but despite this recognition of my condition, I could not, with every exertion of my will, shake off the LSD world. Everything was experienced as completely real, as alarming reality; alarming, because the picture of the other, familiar everyday reality was still fully preserved in the memory for comparison. (20)

As an experiment on the self, the ingestion of LSD-25 was indeed a resounding success—the experimental object was unmistakably transformed, alteration extending even to the agency of Hofmann himself. As a recording, Hofmann could not avoid remembering. For the purposes of LSD-25, Hofmann became what Merlin Donald describes as an "external symbolic storage" device. Combined with Hofmann's own memory of LSD's agency—"And yet I could not forget the relatively uninteresting LSD-25"—Hofmann's transformation by the experiment is subtly but nonetheless inescapably inscribed in the confession: LSD-25 is not easily erased from the experience and memory of the experimental subject, a subject who, like it or not, is recording. Indeed, in some sense Hofmann is both recorder and recording here, as he must respond exegetically to the demands of memory.

Hence while the conclusions to Hofmann's exercise of extreme caution remained to be determined, there could be no argument but that the risk had yielded interesting data, namely, the transformation of Hofmann himself into a being seemingly incapable of forgetting. On this factor alone, Hofmann could determine that he had happened upon a substance of unprecedented potency. Its usefulness and character would, of course, call for further research, but Hofmann's understanding of the causal nature of LSD-25 in his experience was certainly an important result capable of representation to his colleagues.

But if Hofmann had great faith in the splitting capacities of LSD, a molecule that allows for an almost unique position as a retroactive observer and real-time participant, his supervisors, at least at first, did not:

> As expected, the first reaction was incredulous astonishment. Instantly a telephone call came from the management; Professor Stoll asked: "Are you certain you made no mistake in the weighing? Is the stated dose really correct?" Professor

> Rothlin also called, asking the same question. I was certain of this point, for I had executed the weighing and dosage with my own hands. Yet their doubts were justified to some extent, for until then no known substance had displayed even the slightest psychic effect in fraction-of-a-milligram doses. An active compound of such potency seemed almost unbelievable. (21)

While the incredulous questions focused repeatedly on quantitative issues, clearly it was also the qualitative transformations wrought by LSD-25 that inspired disbelief. The missing special report, whatever it says, was evidently unable to carry out its task of translating the delirium of the self-experiment into the allegedly intersubjective space of scientific communication. Hofmann's colleagues were so little convinced by his report that they took the almost unbelievable step of repeating the self-experiment:

> Professor Rothlin himself and two of his colleagues were the first to repeat my experiment, with only one third of the dose I had utilized. But even at that level, the effects were still extremely impressive, and quite fantastic. All doubts about the statements in my report were eliminated. (21)

That is, LSD-25 was in this instance both the object of scientific inquiry and the medium for the communication of the results of that inquiry, the translation of a solo experiment into the general equivalent of truth. While Hofmann's wife returned from Lausanne upon hearing reports that Albert had had some sort of "breakdown," his fellow scientists willingly and immediately ingested LSD in order to eliminate any doubts fostered by Hofmann's report, a psychedelic republic of letters.

It is as an experiment on the self that Hofmann's discoveries are replicated by the community. Only by encountering a veritable undoing of self—a submission to the possible transformation one is in fact testing for—can interesting data from this novel pharmacological agent be gathered, evaluated, and transmitted.

Informatic Prayer

The combination of ineffability common to many mystic traditions and the necessity of communication proper to scientific practice continued to pose problems for the study of hallucinogens as they migrated from Sandoz Pharmaceuticals, where Hofmann worked, to places like the pre-1962 psychology department of Harvard University, the working home of Timothy Leary, Ralph Metzner, Richard Alpert, and Michael Horowitz. Among a parade of intellectuals and artists that moved through Leary's burgeoning circle was writer William S. Burroughs, whose cut-up techniques had, a few years earlier, been used to strip a written text of its meaning—what Burroughs called the virus of the Word—and to allow such texts to interrupt normal consciousness. After a visit with Burroughs, Leary hit upon the idea of using Burroughs's cut-up technique as a framework for reporting the hallucinogenic experience. In this framework, both Burroughs's texts and LSD were experiences that were less understood than undergone and recorded. These recordings—as with Hofmann's precise recall of his LSD trip—were not, however, communicative in the usual sense: they could be understood only retroactively, after the subjects had themselves encountered LSD-25. With the cut-up method, Leary hoped to interrupt the grip of the authorial ego which might interfere with the more direct recording of the LSD experience. In other words, the LSD experience was treated not as an event to be reported on but as an experience to be assayed by writing—the cut-up was a symptom and not a description of the encounter with LSD. In this sense, the information gathered about LSD was understood as a pragmatic rather than semantic production: all data concerned not an understanding of LSD but instead consisted of testing LSD for what it could do. Writing, in this context, becomes less a struggle for intelligibility than a prolix signature of the LSD experience.

Indeed the apparent need to write in response to LSD provoked a graphomania of sorts among Leary's crowd. By 1962 they had been removed from their positions at Harvard University and were now leading itinerant seminars from Zihuatanejo, Mexico, several islands in the Caribbean, and then finally the Hitchcock brothers' mansion in Millbrook, New York. *The Psychedelic Review* regularly published the group's prolific work, and in 1963 Leary, Metzner, and Albert prepared a psychedelic "manual" for use in association with LSD entitled *The Psychedelic Experience*. Published early in 1964, *The Psychedelic Experience* was presented as a source of protocols for the management and study of psychedelic sessions, protocols "based on the Tibetan Book of the Dead." As such, much of the writing was oriented less to "understanding" LSD experience than to dealing with it, enduring it. These texts were intended to be part of the very interface of psychedelia, algorithms for attaining and prolonging particular states: "One may want to pre-record selected passages and simply flick on the recorder when desired. The aim of these instruction texts is always to lead the voyager back to the original First Bardo transcendence and to help maintain that as long as possible."[5]

These recipes and techniques for ecstasy were repeatedly and explicitly linked by Leary to the writing and execution of a sequence of steps in a computer environment: programming.

A third use would be to construct a "program" for a session using passages from

the text. The aim would be to lead the voyager to one of the visions deliberately, or
through a sequence of visions. . . . One can envision a high art of programming
psychedelic sessions, in which symbolic manipulations and presentations would lead
the voyager through ecstatic visionary Bead Games. (98)

In this framework, hallucinogenic subjects become both authors of and platforms
for "symbolic manipulations and presentations," interactive wetware of infinite
experiment and transformation. In the preface to a later work, Leary would write that
"the Psychedelic Experience was our first attempt at session programming."[6]

Crucial to this vision was that the function of information here was to
"program." Less an activity of understanding or even communication than of
repetition and transformation, this program had to be endured. Indeed, the writings
"based" on the Book of the Dead were often not understandable or recognizable until,
like Hofmann's colleagues, one had oneself undergone the encounter with LSD. "The
most important use of this manual is for preparatory reading. Having read the
Tibetan Manual, one can immediately recognize symptoms and experiences which
might otherwise be terrifying, only because of lack of understanding as to what was
happening. Recognition is the key word."[7] Thus the Tibetan Book of the Dead was
treated as a source book, not to be decoded as much as deterritorialized and deployed
in divergent contexts, used more than understood.

Leary made this pragmatic understanding of language (which he saw as
indistinguishable from information) even clearer in *Psychedelic Prayers*, his 1966
"translation" of the *Tao te ching*, what Leary called a "time tested psychedelic manual":

[L]ike all great biblical texts [*sic*], the Tao has been rewritten and re-interpreted
in every century and this is how it should be. The terms for Tao change in each
century. In our times, Einstein rephrases it, quantum theory revises it, the
geneticists translate it in terms of DNA and RNA, but the message is the same.[8]

Translation, for Leary, is both universally available—"the message is the same"—and
utterly variable and reliant upon context, what Leary will refer to as "set and setting"
for the LSD experience but which can be extended to his own work of translation as
well: "these translations from English to psychedelese were made while sitting under
a bamboo tree on a grassy slope of the Kumoan Hills overlooking the snow peaks of
the Himalayas" (38).

Like the drugs and plants with which they are to be used, psychedelic
manuals are catalysts for transcendent experiences—"or can be, given the appropriate
preparation, attitude, and context [the 'set and setting' in Leary's felicitous phrase]"
(10). Crucial to this facility for novel set and settings is not, strictly speaking, its

universal message, but rather its capacity to be translated and travel into novel contexts: "The advice given by the smiling philosophers of China to their emperor can be applied to how to run your home, your office, and how to conduct a psychedelic session" (38).

Note here that the psychedelic manual is not concerned with the communication or an elucidation of a meaning, although it also does so. Rather, the text is seen as a how-to tool for managing and transforming diverse contexts with the help of exotic and yet thoroughly debugged techniques. Its translation is less the production of an equivalent meaning than the "porting" of code to a different platform: "It became apparent that, in order to run exploratory sessions, manuals and programs were necessary to guide subjects through transcendental experiences with a minimum of fear and confusion" (36). This lack of interest in the semantic operation of such manuals is pragmatic: reportage continually fails, even while context is itself a powerful constituent of the psychedelic experience. Hence repetitive failure would be avoided as a negative feedback loop, and other models of abstraction would be experimented with. The incommunicability of the psychedelic experience was taken to be a measure of the complexity of human consciousness as well as the insufficiency of most concepts to it: "No current philosophic or scientific theory was broad enough to handle the potential of a 13 billion-cell computer" (36).

As "prayers," Leary's translations also remind the reader of the active rhetorical register involved in the LSD context: prayers, above all, demand prayer. For Leary, their effects emerge out of their very utterance, a whispered utterance that needs to be said more than heard: "they should be read very slowly and in a serene voice. They should be considered prayers to be whispered" (33). Ideally, then, these rhetorics should approach pure action. Only a trace of the utterance will persist, a whispered scar of language's ecstatic embodiment.

The prayer program was divided by Leary into six parts, and the very center of the sequence is occupied by a twelve-part sutra entitled "Homage to DNA." Here Leary, "translating" and transforming the Book of the Dead into epideictic praise of the book of life, instructs the reader/tripper to "contact cellular consciousness" via the utterance of these meditations. The first, "The Serpent Coil of DNA," invokes a figure also used by C. H. Waddington in his discussion of feedback, the old and now familiar image of the ouroborus:

> [W]e meet it everywhere, but we do not see its front. . . . When we embrace this
> ancient serpent coil, we are masters of the moment, and feel no break in the
> curling back to primeval beginnings. This may be called unraveling the clue of
> the life process (67).

Rather than containing a "message," DNA is hailed as a molecule of ceaseless activity whose very embrace leads to an unraveling or, to translate, undoing. But even as a master of the moment, this "undoing" reveals no central wisdom other than this: our implication in an ancient coil of repetition without beginning or end. Thus rather than triumph, Leary's doggerel suggests an affirmation of primeval complicity, a complicity owed to the very self-replicating ouroborus of DNA, and perhaps, a complicity with LSD and the prayer. The universal message alluded to by Leary is none other than repetition itself.

Contrasted with the molecular biology of the same period—the transmitter of the rhetorics of nucleic acids that were the set and setting of hallucinogenic knowledge—research into psychedelics, both in the lab and in the commune, was strikingly pragmatic in its understanding of information. Both Hofmann and Leary wrote of the need for endurance, a practice that would enable repetition within and beyond the hallucinogenic experience while the very assay of the experience—the self— was becoming variable and even breaking down. Indeed, it was the practice of repetition itself that seems to emerge as crucial to the acquisition and transmission of hallucinogenic knowledges. For Leary and the programmers of psychedelic practice, sessions were sequences of information to be actualized differentially—no two sessions could be enacted in the identical fashion, precisely because of the transformative effects of LSD and its teachings, the repetitive teachings of undoing—"unraveling the clue of the life process." While molecular biology continued to write and talk of decoding the book of life and practicing a hermeneutics of DNA, hallucinogenic discourse suggested that information was less a phenomenon to be understood than it was a potent mutagen of human experience, a mutagen that could only be understood retroactively.

Highway 128, Revisited

If I had not taken LSD ever would I have still been in PCR? I don't know, I doubt it, I seriously doubt it.
(Kary Mullis)

In his 1998 autobiography *Dancing Naked in the Mind Field*, Nobel Prize winner Kary Mullis details his invention of polymerase chain reaction or PCR. While both the topic—the invention of a veritable Xerox machine for nucleic acids—and the genre— scientific autobiography with a hint of scandal, the promise of a forbidden, "naked" truth—encourage the telling of a heroic tale of innovation, Mullis's narrative continually highlights the thoroughly contingent and ungovernable arrival of a concept. Far from simply inflating the role of a lone scientist struggling to know life and the cosmos, *Dancing Naked in the Mind Field* offers a testimony to the thoroughly other, even alien, character of a scientific vision.

By every account, Mullis was himself an unpredictable creature. A biochemist by training, Mullis's most important publication in graduate school treated time reversal and its cosmological implications. In love with the craft of producing new compounds in a more and more efficient fashion, Mullis was every bit as much tinkerer as theorist. This tinkering extended to Mullis's own consciousness. Like many of his U.C. Berkeley chemistry colleagues, Mullis considered the potent new hallucinogens to be experimental objects as real and interesting as any other in their practice. Indeed, after one particularly difficult episode of incorrect dosage, Mullis underwent an entire personality change: overtaken by amnesia, Mullis had the sense that indeed he had become someone else as a result of the encounter.

Nor was Berkeley Mullis's sole connection to hallucinogens. During the period of PCR's emergence as a concept, Mullis visited with Albert Hofmann at the house of his friend Ron Cook. Indeed, Mullis later compares Hofmann's three-lettered discovery with his own:

> The famous chemist Albert Hofmann was at Ron's that night. He had invented LSD in 1943. At the time he didn't realize what he had done. It only slowly dawned on him. And then things worked their way over the years as no one would have predicted, or could have controlled by forethought and reason. Kind of like PCR.[9]

This conjunction of LSD and DNA is of course strikingly resonant with Leary's treatment of both: for Mullis, the main connection between the two was contingency itself, a practice "no one would have predicted, or could have controlled by forethought and reason." Only the unfolding of events allows for a retroactive ascription of narrative and reason—no description of the events could have been compressed into anything shorter than the events themselves. The practice of translating LSD and PCR into their respective technoscientific objects involved nothing less than the transformation of knowledge without precedent, a non sequitur not available to prophecy. There appeared to be no shortcut to either, with only the narrative of accident and coincidence making retroactive sense of the event. Unforeseeable, the effect of both could only be endured, responded to.

But Mullis himself goes further in his account of the connection between PCR and LSD. Retaining the contingency of the concept, Mullis nonetheless attributes to his experience with LSD a certain capacity for the sheer strangeness of the idea of PCR. In a BBC interview, Mullis hesitantly ascribed a paradoxical agency to LSD in the invention of PCR:

> PCR's another place where I was down there with the molecules when I discovered it and I wasn't stoned on LSD, but my mind by then had learned how to get down there. I could sit on a DNA molecule and watch it go by and I didn't feel dumb about that, I felt I could, I mean that's just the way I think is I put myself in all different kind of spots and I've learned that partially I would think, and this is again my opinion, through psychedelic drugs. If you have to think of bizarre things PCR was a bizarre thing. It changed an entire generation of molecular biologists in terms of how they thought about DNA.

Mullis's idea was a remarkably simple one. Indeed, in his cabin off of Highway 128 in Mendocino County, California, the idea struck Mullis as far too simple to be workable: "If the cyclic reactions that by now were symbolized in various ways all over the cabin really worked, why had I never heard of them being used? . . . Why wouldn't these reactions work?"(8–9). Key to Mullis's understanding was the cyclic or iterative character of DNA itself. Unlike earlier molecular biologists who had sought the "meaning" of the genetic code, Mullis sought only to replicate enormous quantities of any given sequence. Mullis writes that because of his knowledge of computer programming, "I understood the power of a reiterative mathematical procedure"(5).

It is precisely as an iteration machine that Mullis treats DNA. While H. G. Khorana and his team had provided the foundation for PCR as early as 1965, they did so within a cryptographic paradigm of discovering the nature and meaning of the genetic code. Francis Crick, writing the lead article for a volume devoted to the triumph of the newly decrypted code, declared that the "historic occasion" was the understanding of the four-letter alphabet of ATCG and its triplet combinations that produce the twenty amino acids, which in turn fold up into proteins: "From all of this we were able to work out the *meaning* of several of the remaining doubtful triplets."[10] Khorana's article appears thirty-six pages later in the volume, and it is clear from its frame and its content that above all Khorana's group sought understanding and even proof: "It was therefore clearly desirable to try to *prove* the total structure of the genetic code by this method."[11] As Paul Rabinow points out in *Making PCR*, the problem of generating DNA sequences was solved by cloning methods, so Khorana's group left the techniques they had discovered for another decade and perspective.[12]

While Rabinow writes that Mullis's approach was the "opposite" of Khorana's, it is perhaps more precise to notice that each had a fundamentally different understanding of information. For Khorana, the genetic code was a space of revelation—the production of a gene was in the service of an understanding of the entire genetic code, an understanding attentive to the rhetorical and epistemological genres of a proof. By contrast, Mullis sought less to understand DNA—for that, after all, was accomplished—than to transform it. It is this pragmatic understanding of genetic information through which PCR emerged.

> Consider, for example, Mullis's account of the PCR process:

> If I could arrange for a short synthetic piece of DNA to find a particular sequence and then start a process whereby that sequence would reproduce itself over and over, then I would be close to solving the problem. . . . The concept was not out of the question because in fact one of the natural functions of DNA molecules is to reproduce themselves.[13]

This treatment of DNA as an ouroborus—not unlike that of Leary's—looked to the pragmatic capacities of nucleic acids: the natural capacity to replicate. And while Rabinow, in his anxiety to avoid a heroic vision of scientific innovation, seeks to distance his own account of PCR from Mullis's, he perhaps overlooks the fact that Mullis was not hero but host, as his instantiation of an iterative, pragmatic understanding of information replicated not just DNA but psychedelic culture itself, as the notion of "expansion" and decontextualization that dislocated consciousness in the 1960s now deterritorialized life, allowing it to become just another sample. Where to begin on the effects of such a deterritorialization or "expansion" of life? Perhaps one must begin with replication. And if any molecule was replicated, perhaps that molecule was LSDNA.

1   Of course, the "code script"—and its subsequent rhetorical transformations of code, program, and network—were all elements of a system to do something: uncover the mysteries of life or, as François Jacob put it, "to triumph over death." In this sense the emphasis on the replication activity of DNA rather than its communicative capacities is the return of the repressed of Schrodinger's machinic and ultimately cryptographic rendering of life. I have discussed this topic in greater length in *On Beyond Living: Rhetorical Transformations of the Life Sciences* (Stanford: Stanford University Press, 1997).

2   Albert Hofmann, *LSD: My Problem Child*, trans. Jonathan Ott (New York: McGraw-Hill, 1980), p. 12.

3   It is not the case, for example, that Hofmann had no options: as his own account details,

many more ways of fathoming LSD-25's effects were available, including the animal testing alluded to earlier.

4   The question of caution is an interesting one here, seemingly oxymoronic to any notion of self-experiment, as the very status of the self is at stake. Hofmann writes that during the terrifying peaks of his hallucinogenic panic, he worried over the possibility of ever communicating the fact of LSD-25's unforeseeable effects to his family. This retroactive recognition of Hofmann's highlights the true experimental conclusion of his work—that there is no preparation, decided or otherwise, for the encounter with the qualitative difference of LSD-25: "Would they ever understand that I had not experimented thoughtlessly, irresponsibly, but rather with the utmost caution, and that such a result was in no way foreseeable? My fear and despair intensified, not only because a young family should lose its father, but also because I dreaded leaving my chemical research work, which meant so much to me, unfinished in the midst of fruitful, promising development. Another reflection took shape, an idea full of bitter irony: if I was now forced to leave this world prematurely, it was because of this lysergic acid diethylamide that I myself had brought forth into the world" (18).

5   Timothy Leary, Ralph Metzner, and Richard Alpert, *The Psychedelic Experience: A Manual Based on the Tibetan Book of the Dead* (New Hyde Park, New York: University Books, 1964), p. 97.

6   Timothy Leary, *Psychedelic Prayers and Other Meditations* (Berkeley: Ronin, 1997), p. 37.

7   Leary, *The Psychedelic Experience*, p. 97.

8   Leary, *Psychedelic Prayers*, p. 37.

9   Kary Mullis, *Dancing Naked in the Mind Field* (New York: Pantheon Books, 1998), p. 11.

10   F. H. C. Crick, "The Genetic Code: Yesterday, Today, and Tomorrow," *Cold Spring Harbor Symposia on Quantitative Biology* 31 (1966): 3. Emphasis added.

11   Ibid., p. 39. Emphasis added.

12   Paul Rabinow, *Making PCR: A Story of Biotechnology* (Chicago: University of Chicago Press, 1996).

13   Mullis, *Dancing Naked*, p. 6.

Doyle's "LSDNA"

Steven Shaviro

Richard Doyle's essay makes provocative connections between the biotechnology that is increasingly reshaping our world today, on the one hand, and the illicit "technologies of the sacred" associated with LSD and other psychedelic drugs, on the other. These connections work on a number of levels. Psychedelic research and genetic engineering are both exercises in applied biochemistry. LSD has had a direct influence on important advances in biotechnology, most notably Kary Mullis's invention of the polymerase chain reaction. Early LSD researchers like Timothy Leary and his associates made extensive, explicit use of figures and paradigms drawn from the contemporary understanding of DNA; conversely, DNA research seems to make at least an implicit use of procedures and assumptions earlier found in LSD study. More deeply, DNA research and LSD research share common presuppositions and goals. Both are technologies that seek to "reprogram" the human body (itself an interesting metaphor, since it implies that organic processes are analogous to the operations of digital computers). Both thereby raise fundamental questions of identity and agency. And, most importantly, both types of research manifest the same shift from hermeneutics to pragmatics: from an attempt to decipher hidden meanings, to an attempt to produce visible and repeatable effects.

These analogical and genealogical connections are disconcerting, in part because of the precisely reciprocal positions that the two sorts of research in question occupy in the world (and particularly in America) today. Governments and corporations are investing significant resources in biotechnological research; while a sizeable, passionate minority, centered in what might still be called (with reservations) the counterculture, finds this research dangerous and abhorrent, and has even engaged in terrorist activities in order to stop it. Conversely, a sizeable and passionate countercultural minority has been increasingly active over the last decade in secret, unauthorized psychedelic experimentation; while governments, and society in general, continue to find this activity dangerous and abhorrent, and have tried to stop it with severe legal sanctions. Consider, too, the historical irony that psychedelic research was originally promoted by the United States government, under the auspices of the Department of Defense and the CIA. Meanwhile, the threat of illicit, potentially catastrophic bioengineering, performed by political terrorists or mafias and drug cartels, is rapidly crossing over from science fiction to actual fact. This mirror relation between the uses of DNA and LSD is one of those all-too-neat binary oppositions that just cries out for the sort of subversive deconstruction and reinscription that Doyle gives it here.

Both of the technologies discussed by Doyle have already had an irreversible effect upon the world we live in. Genetically engineered crops have long since emerged out of the labs and into the farms and fields; they are now, like it or not, part of "nature." As for LSD and other vehicles of psychedelic experience, it's true that the pursuit of ecstasy and insight through chemical means has been harshly repressed; but at the same time, we have seen the massive spread of legal chemicals, like Prozac, Zoloft, and Paxil, which target the same 5-HT2A serotonin receptors as LSD, and which are prescribed for not entirely unrelated therapeutic ends. Beyond these cases, it is quite likely that, despite efforts to stop them, both gene-altering and serotonin-affecting technologies will be even more influential in the years to come. With LSD as with DNA, there's no way to put the genie back in the bottle.

Doyle's surprising juxtaposition of biotechnology and psychedelic technology brings into sharp relief certain features of both that might not have been noticed otherwise. One of these is the idea of self-experimentation. Recent mainstream accounts of the uses and abuses of illicit drugs tend to say, somewhat disparagingly, that incidents like Dr. Albert Hofmann's first LSD trip, which do not fit into orthodox research protocols, date from "the era when scientific self-experimentation was more common than it is today."[1] Of course this is rather disingenuous, since the real reason that such experiments are less prevalent today is that they have been driven underground. Doyle shows, however, that the kind of self-experimentation that violates traditional norms of scientific objectivity is structurally necessary to the very project of psychedelic research. This is because the "objective" effect of psychedelic drugs is to transform irreversibly the very subject who has been examining them. I am no longer the same person I was before the experiment began. The experience of LSD short-circuits the distinction between the observing self (the self as transcendental subject) and the self being observed (the self as an object in the world). But since such a "split subject"[2] or "empirico-transcendental doublet"[3] is central to the very project of modern scientific reason, its disruption has the effect of a radical mutation in the space of our understanding. And, despite its surface conformity to scientific protocols, genetic engineering partakes also, at least potentially, of this disturbing power: it isn't just something that human beings do, but something that tends to turn us into something else, something posthuman or transhuman.

This leads directly to the second crucial characteristic of both genetic and psychedelic experimentation, as described by Doyle: they have increasingly come to be pragmatic exercises, rather than interpretive ones. They are not oriented toward understanding the world, but toward changing it. They even seem to work in such a manner that they permit us (or better, incite us) to change the world, precisely to the extent that they prevent us from understanding it. The promise, as well as the danger, of both genetic and psychedelic engineering is that they bind us to an unforeseeable futurity. Gene-splicing and consciousness-altering activities compel us to concede more to the biochemical realm than we might otherwise want to (since they wreak havoc upon our ideas of free will and responsibility) and yet to reject any sort of biological determinism (since they teach us how to induce biochemical changes at will—even if we are not in control of the consequences). This is how genetics and psychedelics alike put our very being into question.

1   Cynthia Kuhn, Scott Swartzwelder, and Wilkie Wilson, *Buzzed: The Straight Facts about the Most Used and Abused Drugs* (New York: W. W. Norton, 1998), pp. 83–84.

2   Jacques Lacan, *The Four Fundamental Concepts of Psycho-Analysis*, trans. Alan Sheridan (New York: W. W. Norton, 1978), p. 138 ff.

3   Michel Foucault, *The Order of Things: An Archaeology of the Human Sciences*, trans. not credited (New York: Vintage Books, 1970), p. 318 ff.

The Virtual Surgeon:
Operating on the Data in an Age of Medialization

Timothy Lenoir

Media inscribe our situation. We are becoming immersed in a growing repertoire of computer-based media for creating, distributing, and interacting with digitized versions of the world, media that constitute the instrumentarium of a new epistemic regime. In numerous areas of our daily activities, we are witnessing a drive toward the fusion of digital and physical reality; not the replacement of the real by a hyperreal, the obliteration of a referent and its replacement by a model without origin or reality as Baudrillard predicted, but a new playing field of ubiquitous computing in which wearable computers, independent computational agent-artifacts, and material objects are all part of the landscape. To paraphrase William Gibson's character Case in *Neuromancer*, "data is being made flesh."

Surgery provides a dramatic example of a field newly saturated with information technologies. In the past decade, computers have entered the operating room to assist physicians in realizing a dream they've pursued ever since Claude Bernard: to make medicine both experimental and predictive. The emerging field of computer-assisted surgery offers a dramatic change from the days of individual heroic surgeons. Soon surgeons will no longer boldly improvise on modestly preplanned scripts, adjusting them in the operating room to fit the peculiar case at hand. To perform an operation, surgeons must increasingly use extensive three-dimensional-modeling tools to generate a predictive model, the basis for a simulation that will become a software-surgical interface. This interface will guide the surgeon in performing the procedure.

The Minimally Invasive Surgery Revolution

These developments in surgery date back to the 1970s when widely successful endo-scopic devices appeared. First among these were arthroscopes for orthopedic surgery, available in most large hospitals by 1975, but at that point endoscopy was more a gimmick than a mainstream procedure. Safe surgical procedures with such scopes were limited because the surgeon had to operate while holding the scope in one hand and a single instrument in the other.

What changed the image of endoscopy in the mind of the surgical commu-nity and turned arthroscopy, cholecystectomy (removal of the gallbladder with instruments inserted through the abdominal wall), and numerous other endoscopic surgical techniques into common operative procedures? The introduction of the small medical video camera attachable to the eyepiece of the arthroscope or laparoscope was an initial major step. French surgeons were the first to develop small, sterilizable, high-resolution video cameras that could be attached to a laparoscopic device. With the further addition of halogen high-intensity light sources with fiber-optic connec-tions, surgeons were able to obtain bright, magnified images that could be viewed on a video monitor by all members of the surgical team rather than by just the surgeon alone. This technical development had consequences for the culture of surgery; it contributed to greater cooperative teamwork and opened the possibility for surgical procedures of increasing complexity, including suturing and surgical reconstruction done only with videoendoscopic vision.[1] French surgeons performed the first laparoscopic cholecystectomy in 1989. A burgeoning industry in biomedical devices, such as new, specialized instruments for tissue handling, cutting, hemostasis, and more, sprang up almost immediately to provide the necessary ancillary technology to make laparoscopic procedures practical in your local hospital.

Due to their benefits of small scars, less pain, and a more rapid recovery, endoscopic procedures were rapidly adopted after the late 1980s and became a standard method for nearly every area of surgery in the 1990s. Demand from patients has had much to do with the rapid evolution of the technology. Equally important have been the efforts of health care organizations to control costs. In a period of deep concern about skyrocketing health care costs, any procedure that improved surgical outcomes and reduced hospital stays interested medical-instrument makers. Encouraged by the success of the new videoendoscopic devices, medical-instrument companies in the early 1990s foresaw a new field of minimally invasive diagnostic and surgical tools. Surgery was about to enter a technology-intense era that offered immense opportunities to companies teaming surgeons and engineers to apply the latest developments in robotics, imaging, and sensing to the field of minimally invasive surgery. While pathbreaking developments had occurred, the instruments available for such surgeries allowed only a limited number of the complex functions demanded by the surgeon. Surgeons needed better visualization, finer manipulators, and new types of remote sensors, and they needed these tools integrated into a complete system.

Telepresence Surgery

A new vision emerged, heavily nurtured by funds from the Advanced Research Projects Agency (ARPA), the NIH, and NASA, and developed through contracts made by these agencies to laboratories such as the Stanford Research Institute (SRI), the Johns Hopkins Institute for Information Enhanced Medicine, the University of North Carolina Computer Science Department, the University of Washington Human Interface Technology Laboratory, the Mayo Clinic, and the MIT Artificial Intelligence Laboratory. The vision promoted by Dr. Richard Satava, who spearheaded the ARPA program, was to develop "telepresence" workstations that would allow surgeons to telerobotically perform complex surgical procedures that demand great dexterity. These workstations would re-create and magnify all of the motor, visual, and tactile sensations the surgeon would actually experience inside the patient. The aim of the programs sponsored by these agencies was eventually to enable surgeons to perform surgeries, such as certain complex brain surgeries or heart operations not even possible in the early 1990s, improve the speed and surety of existing procedures, and reduce the number of people in the surgical team. Central to this program was telepresence-telerobotics, allowing operators the complex sensory feedback and motor control they would have if they were actually at the work site, carrying out the operation with their own hands. The goal of telepresence was to project full motor and sensory capabilities—visual, tactile, force, auditory—into even microscopic environments to perform operations that demand fine dexterity and hand-eye coordination.

Philip Green led a team at SRI that assembled the first working model of a telepresence surgery system in 1991, and with funding from the NIH Green went on to design and build a demonstration system. The proposal contained a diagram showing the concept of workstation, viewing arrangement, and manipulation configuration used in the surgical telepresence systems today (fig. 1). In 1992 SRI obtained funding for a second-generation telepresence system for emergency surgeries in battlefield situations. For this second-generation system, the SRI team developed the precise servo-mechanics, force-feedback, three-dimensional visualization, and surgical instruments needed to build a computer-driven system that could accurately reproduce a surgeon's hand motions with remote surgical instruments having five degrees of freedom and extremely sensitive tactile response (fig. 2).

In late 1995 SRI licensed this technology to Intuitive Surgical, Inc., of Mountain View, California. Intuitive Surgical furthered the work begun at SRI by improving on the precise control of the surgical instruments, adding a new invention, EndoWrist, patented by company cofounder Frederic Moll, which added two degrees of freedom to the SRI device—inner pitch and inner yaw (inner pitch is the motion a wrist performs to knock on a door; inner yaw is the side-to-side movement used in wiping a table)—allowing the system to better mimic a surgeon's actions; it gives the robot the ability to reach around, beyond, and behind delicate body structures, delivering these angles right at the surgical site. Through licenses of IBM patents, Intuitive also improved the three-dimensional video imaging, navigation, and registration of the video image to the spatial frame in which the robot operates. The system employs 250 megaflops of parallel processing power (figs. 3, 4).

A further crucial improvement to the system was brought by Kenneth Salisbury from the MIT Artificial Intelligence Laboratory. Salisbury imported ideas from the force-reflecting haptic feedback system he and Thomas Massie invented as the basis of their PHANTOM system,[2] a device invented in 1993 permitting touch interactions between a human user and a remote virtual and physical environment. The PHANTOM is a desktop device that provides a force-reflecting interface between the user and a computer. Users connect to the mechanism by simply inserting their index finger into a thimble. The PHANTOM tracks the motion of the user's fingertip and can actively exert an external force on the finger, creating compelling illusions of interaction with solid physical objects. A stylus can be substituted for the thimble and users can feel the tip of the stylus touch virtual surfaces.

The haptic interface allows the system to go beyond previous instruments for minimally invasive surgery (MIS). These earlier instruments precluded a sense of touch or feeling for the surgeon; the PHANTOM haptic interface, by contrast, gives an additional element of immersion. When the arm encounters resistance inside the patient, that resistance is transmitted back to the console, where the surgeon can feel it. When the thimble hits a position corresponding to the surface of a virtual object in the computer, three motors generate forces on the thimble that imitate the feel of the

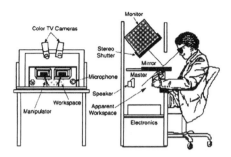

Figure 1.
Philip Green, schema for force-reflecting
surgical manipulator, Stanford Research
Institute, Menlo Park, CA, 1992

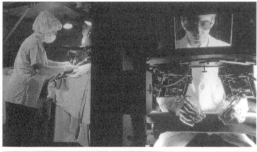

Figure 2.
Philip Green, force-reflecting surgical
manipulator, *Time* magazine, Special Issue,
Fall 1996

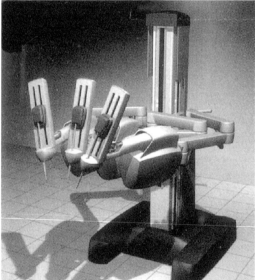

Figure 3.
Intuitive Surgical DaVinci Computer
Assisted Robotic Unit, from Intuitive
Surgical promotional material, Intuitive
Surgical, Palo Alto, CA, 1999

Figure 4.
Endoscopic bypass surgery, Paris, 1999,
using Intuitive Surgical system, photo
from Intuitive Surgical press release

object. The PHANTOM can duplicate all sorts of textures, including coarse, slippery, spongy, or even sticky surfaces. It also reproduces friction. And if two PHANTOMs are put together a user can "grab" a virtual object with thumb and forefinger. Given advanced haptic and visual feedback, the system greatly facilitates dissecting, cutting, suturing, and other surgical procedures, even those on very small structures, by giving the doctor inches to move in order to cut millimeters. Furthermore, it can be programmed to compensate for error and natural hand tremors that would otherwise negatively affect MIS technique.

The surgical manipulator made its first public debut in actual surgery in May of 1998. From May through December 1998, Professor Alain Carpentier and Dr. Didier Loulmet of the Broussais Hospital in Paris performed six open-heart surgeries using the Intuitive system.[3] In June of 1998, the same team performed the world's first closed-chest videoendoscopic coronary bypass surgery completely through small (1 cm) ports in the chest wall. Since that time more than 250 heart surgeries and 150 completely videoendoscopic surgeries have been performed with the system. The system was given approval to be sold throughout the European Community in January of 1999.

Computer Modeling and Predictive Medicine

A development of equal importance to the contribution of computers in the MIS revolution has been the application of computer modeling, simulation, and virtual reality to surgery. The development of various modes of digital imaging in the 1970s, such as CT (which was especially useful for bone), MRI (useful for soft tissue), ultrasound, and later PET scanning have made it possible to do precise quantitative modeling and preoperative planning for many types of surgery. Because these modalities, particularly CT and MRI, produce two-dimensional "slices" through the patient, the natural next step (taken by Gabor Herman and his associates in 1977) was to stack these slices in a computer program to produce a three-dimensional visualization.[4] Three-dimensional modeling first developed in craniofacial surgery because it focused on bone, and CT scanning was more highly evolved. Another reason was that in contrast to many areas of surgery where a series of two-dimensional slices—the outline of a tumor for example—provides all the information the surgeon needs, in craniofacial surgery the surgeon must focus on the skull in its entirety rather than on one small section at a time.

Jeffrey March and Michael Vannier pioneered the application of three-dimensional computer imaging to craniofacial surgery in 1983.[5] Prior to their work, surgical procedures were planned with tracings made on paper from two-dimensional radiographs. Frontal and lateral radiographs were taken and the silhouette lines of bony skull edges were traced onto paper. Cutouts were then made of the desired bone fragments and manipulated. The clinician would move the bone fragment cutout in the paper simulation until the overall structure approximated normal. Measurements would be taken and compared to an ideal, and another cycle of cut-and-try would be carried out. These hand-done optimization procedures would be repeated until a surgical plan was derived that promised to yield the most normal-looking face for the patient.

Between 1983 and 1986, March, Vannier, and their colleagues computerized each step of this two-dimensional optimization cycle.[6] The three-dimensional visualizations overcame some of the deficiencies in the older two-dimensional process. Two-dimensional planning is of little use in attempting to consider the result of rotations. Cutouts planned in one view are no longer correct when rotated to another view. Volume rendering of two-dimensional slices in the computer overcame this problem. Moreover, comparison of the three-dimensional preoperative and postoperative visualizations often suggested an improved surgical design in retrospect. A frequent problem in craniofacial surgery is the necessity of having to perform additional surgeries to get the optimal final result. For instance, placement of bone grafts in gaps leads to varying degrees of resorption. Similarly, a section of the patient's facial bones may not grow after the operation, or attachment of soft tissues to bone fragments may constrain the fragments' movement. These and other problems suggested the value of a surgical simulator that would assemble a three-dimensional interactive model of the patient from imaging data, provide the surgeon with tools similar to engineering computer-aided design tools for manipulating objects, and allow him or her to compare "before" and "after" views to generate an optimal surgical plan. In 1986 March and Vannier developed the first simulator by using commercial CAD software to provide an automated optimization of bone fragment position to "best fit" normal form.[7] Since then, customized programs designed specifically for craniofacial surgery have made it possible to construct multiple preoperative surgical plans for correcting a particular problem, allowing the surgeon to make the optimal choice.

These early models were further extended in an attempt to make them reflect not only the geometry but also the physical properties of bone and tissues, thus rendering them truly quantitative and predictive. R. M. Koch, M. H. Gross, and colleagues from the ETH (Eidgenössische Technische Hochschule) Zürich, for example, applied physics-based finite element modeling to facial reconstructive surgery.[8] Going beyond a "best fit" geometrical modeling among facial bones, their approach is to construct triangular prism elements consisting of five layers of epidermis, dermis, subcutaneous connective tissue, fascia, and muscles, each connected to one another by springs of various stiffnesses. The stiffness parameters for the soft tissues are assigned on the basis of segmentation of CT scan data. In this model each prism-shaped volume element has its own physics. All interactive procedures, such as bone and soft-tissue repositioning, are performed under the guidance of the modeling system, which feeds the processed geometry into the finite element modeling program. The resulting shape is generated by minimizing the global energy of the surface under the presence of external forces. The result is the ability to generate highly realistic three-dimensional images of the postsurgical shape. Computationally based surgery analogous to the craniofacial surgery described above has been introduced in eye surgeries, in prostate, orthopedic, lung, and liver surgeries, and in repair of cerebral aneurysms.

Equally impressive applications of computational modeling have been introduced into cardiovascular surgery. In this field, simulation techniques have gone beyond modeling structure to simulating function, such as blood flow in the individual patient who needs, for example, coronary bypass surgery. Charles A. Taylor and colleagues at the Stanford Medical Center have demonstrated a system that creates a patient-specific three-dimensional finite element model of the patient's vasculature and blood flow under a variety of conditions.[9] A software simulation system using equations governing blood flow in arteries then provides a set of tools that allows the physician to predict the outcome of alternate treatment plans on vascular hemodynamics. With such systems, predictive medicine has arrived.

Medical Avatars: Surgery as Interface Problem

Such examples demonstrate that computational modeling has added an entirely new dimension to surgery. For the first time, the surgeon is able to plan and simulate a surgery based on a mathematical model that reflects the actual anatomy and physiology of the individual patient. Moreover, the model need not stay outside the operating room. Several groups of researchers have used these models to develop "augmented reality" systems that produce a precise, scaleable registration of the model on the patient so that a fusion of the model and the three-dimensional stereo camera images is made. The structures rendered from preoperative MRI or CT data are registered on the patient's body and displayed simultaneously to the surgeon in near-to-real-time. Intense efforts are underway to develop real-time volume rendering of CT, MRI, and ultrasound data as the visual component in image-guided surgery. Intraoperative position-sensing enhances the surgeon's ability to execute a surgical plan based on three-dimensional CT and MRI by providing a precise determination of his tools' locations in the geography of the patient. This procedure has been carried out successfully in removing brain tumors and in a number of prostatectomies in the Mayo Clinic's Virtual Reality Assisted Surgery Program (VRASP) headed by Richard Robb.

In addition to improving the performance of surgeons by putting predictive modeling and mathematically precise planning at their disposal, computers are playing a major role in improving surgical outcomes by providing surgeons opportunities to train and rehearse important procedures before they go into the operating theater. By 1995, modeling and planning systems began to be implemented in both surgical training simulators and in real-time surgeries. One of the first systems to incorporate all these features in a surgical simulator was developed for eye surgery by MIT robotics scientist Ian Hunter (fig. 5). Hunter's microsurgical robot (MSR)

Figure 5.
Ian Hunter's microsurgical robot, *Presence:
Teleoperators and Virtual Environments*,
vol. 2, 1993

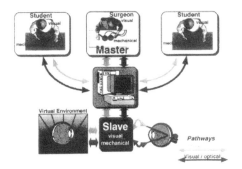

Figure 6.
Ian Hunter et al., *Presence*, vol. 2, 1993,
showing "fade-in" of student surgeons

system incorporated features described above, such as data acquisition by CT and MRI scanning, use of finite element modeling of the planned surgical procedure, a force-reflecting haptic feedback system which enables the perception of tissue-cutting forces, including those that would normally be imperceptible if they were transmitted directly to the surgeon's hands.[10]

Surgery demands an interface. The surgeon is on the outside. The targeted anatomy is on the inside. Minimally invasive laparoscopic surgery is typically performed by making a small incision in the patient's body and inserting a long shafted instrument. At the far end of the shaft is the working tip of the instrument that contacts the target anatomy inside the patient. At the near end of the shaft is the mechanism (typically finger loops) handled by the surgeon outside the patient. The mechanism outside the patient is the master component that controls the action of the slave mechanism inside the patient. The shaft provides a physical link or interface between master and slave. But laparoscopic systems have a number of problems. While minimally invasive laparoscopic surgical methods permit smaller entry incisions, the entry point fulcrum inverts hand movements, limits degrees of freedom, and amplifies tremor, making the surgery more difficult.

Robotic systems combining virtual reality interfaces with haptic feedback, such as Hunter's prototype and a similar system developed by researchers at the University of Washington's Human Interface Technology Laboratory (HIT Lab), can overcome these problems with minimally invasive laparoscopic methods.[11] By performing the procedure with a robot, one can numerically remap the relationship between the surgeon and the instruments. The surgeon's head and hand movements are tracked by the system. The system performs inverse kinematic transformations so that the artifacts of the fulcrum point are effectively bypassed, making the surgeon's movements appear to drive the instruments as if he or she were literally present at the site of the surgical procedure. This provides more direct manipulations resembling those of open surgery, while maintaining the benefits of minimal incision. By controlling the articulated endoscope with the surgeon's head movements and feeding the endoscopic image back to a head-mounted display, the system gives the surgeon the impression of being immersed into the patient's body. Additional scaling transformations and tremor-filtering convert large movements by the surgeon into smoothed, accurate, microsurgical movements by the robot.

Immersive robotic surgical interfaces fusing the haptic environment with three-dimensional stereo camera images fed to a head-mounted display give the surgeon the perspective of being placed inside the patient's body and shrunk to the scale of the target anatomy. Such systems are valuable as training devices. As if in a flight simulator, the surgeon can rehearse the procedure on a model of the individual patient. In addition, the model can be used as a training site for student surgeons, present during a practice surgery, sharing the same video screen and feeling the same surgical moves as the master surgeon. Such systems can also be deployed in a collaborative telesurgery system, allowing different specialists to be faded in to "take the controls" during different parts of the procedure (fig. 6). Indeed, a "collaborative clinic" incorporating these features was demonstrated at NASA-Ames on May 5, 1999, with participants at five different sites around the United States.

Such demonstrations point to the possibility in the not distant future of a new type of operating theater. In place of the typical scene of the crowded operating theater with assistants and technicians, we could expect to see a lone surgeon seated at an operating console powered by Silicon Graphics Infinite Reality Engines, communicating simultaneously with participant surgeons located at distant sites, with online access to virtual reference tools including a library of distributed virtual objects and the databanks of the National Institutes of Health's Digital Human via the Scaleable Coherent Interface on Fiber Channel at eight gigabits per second. Although seated alone at his console, the surgeon would actually be assisted by a team of surgeons and support technicians with whom he is virtually present in an operating room; they see him as he performs the delicate surgery with them.

A scenario projected five to ten years into the future by the National Research Council's Committee on Virtual Reality Research illustrates how future surgeons may be trained to use these surgical interfaces. In a discussion of the use of VR in training heart surgeons, VR researchers describe how haptic augmentation can correct the tremors of the hand as it guides a scalpel over a beating heart:

> Jennifer Roberts . . . is training to become a surgeon and is at her SE (surgical environment) station studying past heart operations. . . . This system includes a special virtual-heart computer program obtained from the National Medical Library of Physical/Computational Models of Human Body Systems and a special haptic interface that enables her to interact manually with the virtual heart. Special scientific visualization subroutines enable her to see, hear, and feel the heart (and its various component subsystems) from various vantage points and at various scales. Also, the haptic interface, which includes a special suite of surgical tool handles for use in surgical simulation (analogous to the force-feedback controls used in advanced simulations of flying or driving), enables her to practice various types of surgical operations on the heart. As part of this practice, she sometimes deliberately deviates from the recommended surgical procedures in order to

observe the effects of such deviations. However, in order to prevent her medical school tutor (who has access to stored versions of these practice runs on his own SE station) from thinking that these deviations are unintentional (and therefore that she is poor material for surgical training), she always indicates her intention to deviate at the beginning of the surgical run.

　　Her training also includes studying heart action in real humans by using see-through displays (augmented reality) that enable the viewer to combine normal visual images of the subject with images of the beating heart derived (in real time) from ultrasound scans.

　　. . . In all of these operations, the surgery was performed by means of a surgical teleoperator system. Such systems not only enable remote surgery to be performed, but also increase surgical precision (e.g., elimination of hand tremor) and decrease need for immobilization of the heart during surgery (the surgical telerobot is designed to track the motion of the heart and to move the scalpel along with the heart in such a way that the relative position of the scalpel and the target can be precisely controlled even when the heart is beating).

　　The human operator of these surgical teleoperator systems generally has access not only to real-time visual images of the heart via the telerobotic cameras employed in the system, but also to augmented-reality information derived from other forms of sensing and overlaid on the real images. Some of these other images, like the ultrasound image mentioned above, are derived in real time; others summarize information obtained at previous times and contribute to the surgeon's awareness of the patient's heart history.

　　All the operations performed with such telerobotic surgery systems are recorded and stored using visual, auditory, and mechanical recording and storage systems. These operations can then be replayed at any time (and the operation felt as well as seen and heard) by any individual such as Jennifer, who has the appropriate replay equipment available. Recordings are generally labeled "master," "ordinary," and "botched," according to the quality of the operation performed. As one might expect, the American Medical Association initially objected to the recording of operations; however, they agreed to it when a system was developed that guaranteed anonymity of the surgeon and the Supreme Court ruled that patients and insurance companies would not have access to the information. This particular evening, Jennifer is examining two master double-bypass operations and one botched triple-bypass operation.[12]

This scenario builds its vision of the future from systems like Hunter's microsurgical robot. Among the many remarkable features in this account, perhaps one of the most salient for my purposes is the medialization and simultaneous rewriting of human agency depicted. The Committee on Virtual Reality Research focuses on the utility of the system for teaching purposes. In Hunter's system, multiple participants can be "faded in" and "faded out" so that they actually feel what the surgeon directing the robot feels. But here a reverse video effect seems to set in: it is difficult to determine who is in control, robot system or human. A human team clearly programs the robot, but the robot enhances perception and actually guides the hand of the surgeon, correcting for errors due to (human-generated) hand tremor. The guiding hand of the microsurgical system "trains" Jennifer's erratic movements.

Surgery in an Age of Medialization

The microsurgical systems I have sketched above are by no means wild fantasies of techno-enthusiast surgeons. After little more than a decade of serious development, many of these systems are already in use in select areas in Europe, and several have been approved for clinical trials in the United States. To be sure, these developments do not represent a large movement in contemporary medicine; they account for a fraction of the funds spent on medical development. Nevertheless, it is intriguing to ponder the conditions that would lead them to be implemented more extensively and the consequences entailed for both patients and surgeons were these technologies to become widely adopted. Let's begin by considering the arguments of proponents of the systems and the economic and political pressures that support their efforts.

Advocates of these systems claim that cost savings will result from the new technologies. Surgeries that are more accurately planned, less invasive, and more precisely executed can reduce blood loss and improve patients' recovery rates. Proponents also point to more efficient use of costly facilities through telepresence and the improvement of training regimes for surgeons. Such arguments question our tolerance for high error rates in surgeries (greater than 10 percent in some areas) whereas in other areas of risk, such as pilot training for commercial airlines, we would find even a 2 percent error rate intolerable. In the case of pilot error, one reason for the low incidence of error is arguably the availability of high-quality simulation technology for training.

A salient feature of contemporary health care is its attention to designing health care plans: diagnoses and therapies targeted for the individual patient. This coincides with the demand for greater involvement by patients in decisions related to their own health. The new surgical techniques map onto the preference for individually tailored therapies. As I have suggested above, the new modeling and simulation tools allow procedures to be designed on the basis of actual patient data rather than on generic experience with a condition—procedure x is what you do in situation y. Dynamic simulation and modeling tools enable surgeons to construct alternative surgical plans using actual anatomic and physiological data projected to specific outcomes in terms of lifestyle and patient expectations. Proponents argue that the new surgical tools take the guesswork out of choosing a procedure tailored to the case at hand. Such outcomes not only increase patient satisfaction but reduce costly repetition of procedures that were not optimized on the first pass.

The downside of this greater precision for the patient, of course, is increased surveillance. It is strangely ironic that while the new technology brings the capability to design therapies—including drugs—specifically targeted for the individual, and hence freeing the individual from infirmity and disease in a way never before imagined, it does so most efficiently and cost-effectively by instituting a massive system of preventive health care from genome to lifestyle. In the age of medialization, your lifestyle is medicalized.

It is not difficult to see how the surgical technologies explored here would mesh with such a system. They deploy anatomical overlays and patient-related data as aids to the surgical procedure, but other layers of augmentation can be foreseen. Analogous to the inclusion of material constraints, cost-factors, and building-code regulations in current CAD-CAM design tools, surgical simulators could be augmented with a list of procedures authorized by the patient's HMO, and within this list various treatment packages could be prescribed according to the benefit plan. In a number of states, hospitals and managed care facilities that receive reimbursement from Medicaid are required to treat patients with a prioritized list of diagnoses and procedures, ranked according to criteria such as life expectancy, quality of life, cost effectiveness of a treatment, and the scope of its benefits. The Oregon Health Plan, which first implemented this system, ranked seven hundred diagnoses and treatments in order of importance. Items below line 587 are disallowed.[13] Currently in facilities such as emergency rooms, a staff supervisor examines the treatment prescribed by staff physicians. The prescribing physician must produce formal written justification in support of any decision to ignore the guidelines. Physicians are reluctant to confront this additional layer of bureaucracy, particularly since the financial risks incurred by denial of Medicaid funding can be a potential source of friction with the management of the HMO employing them. In the future, the appropriate constraints and efficiency measures could be preprogrammed into the surgical treatment planning simulator.

The new computer-intensive, highly networked surgical systems I have explored also carry consequences for the discipline of surgery and for the agent we call "surgeon." In the age of heroic medicine, before the advent of the corporate health care system, surgeons were celebrated as among the most autonomous of professional agents. Society granted these demigods of the surgical wards great status and autonomy in exchange for their ability to bring massive amounts of scientific and medical knowledge to bear in a heartbeat of surgical practice.[14] These "guys" (since surgeons were overwhelmingly males) had the proverbial "right stuff," agency par excellence. But in the telerobotics systems examined here, the surgeon-function dissolves into the ever more computationally mediated technologies of apperception, diagnosis, decision, gesture, and speech. The once autonomous surgeon agent is being displaced by a collection of software agents embedded in megabits of computer code. How is this possible?

Consider the surgeon planning an arterial stent graft before the advent of real-time volume rendering. A medical atlas—or perhaps more recently a three-dimensional medical viewer—was used in combination with echocardiograms, CT scans, and MRI images of the patient. At best the surgeon dealt with a stack of two-dimensional representations, slices separated by several millimeters. These were mentally integrated in the surgeon's imagination and compared with the anatomy of the standard human. Through this complex process of internalization, reasoning, and imagining, surgeons "saw" structures they would expect to be seeing as they performed the actual surgery, a quasi-virtual surgical template in their imagination. The surgeon worked as the head of a team in the operating room, with anesthesiologists and several surgical assistants, but it was the surgeon as an individual who mentally planned and executed the surgery. No matter how you slice it, the position of the surgeon as an autonomous center of agency and responsibility was crucial to this system.

In the new surgical paradigm, the surgeon first begins with the patient dataset of MRI, CT, and other physiological data. He or she enters that data into a surgical model utilizing a variety of software and data management tools to construct a simulation of the surgery to be performed. The Virtual Workbench, Cyberscalpel, and various systems for interfacing anatomical and physiological data with finite element modeling tools are all elements of this new repertoire of tools for preparing a surgery. A surgical plan is constructed listing the navigational coordinates, step-by-step procedures, and specific patient data important to keep in mind at critical points. The simulation is, in fact, an interactive hypermedia document.

Voxel-Man, a virtual, three-dimensional atlas of anatomy, provides a particularly clear illustration of this hypertextualization of the surgical body.[15] Its approach is to combine in a single framework a computer-generated spatial model and an atlas (containing textual descriptions) of the details of every volume element in the anatomical structures along the path of the surgery. These constituents vary with the different domains of knowledge, such as structural and functional anatomy. The same voxel (volume pixel element) may belong to different voxel sets with respect to the particular domain. The membership is characterized by object labels that are stored in "attribute volumes" congruent to the image volume, including features like vulnerability or mechanical properties, which might be important for the surgical simulation. Also included can be patient-specific data for that particular region, such as the specific frames of MRI or CT data used to construct the simulation.

Such intelligent volumes are not only for preparing the surgery, or later for teaching and review. Built into the patient-specific surgical plan, the hypertext atlas assumes the role of surgical companion in an "augmented reality" system. In Hunter's surgical manipulator, for example, various pieces of information—patient-specific data such as MRI records, or particular annotations the surgical team had made in preparing the plan—appear in the margins of the visual simulation indicating particular aspects of the procedure to be performed at a given stage of the surgery. The surgeon-team and the procedures it designs are thus inscribed in a vast hypertext narrative of spatialized scripts to be activated as the procedure unfolds.

Well before we enter the operating room of the future, it is clear that surgeons will be significantly reconfigured in terms of skills and background. Two processes are driving that reconfiguration: medialization and postmodern distributed production. Key to medialization is the externalization of formerly internal mental processes, the literalization of skill in an inscription device.[16] This process is abundantly evident in the introduction of new media technologies in surgery, such as computer visualization, modeling and simulation modules, and computer-generated virtual reality interfaces for interacting with the patient's body. Whereas various aspects of the visualization and presurgical planning took place in the surgeon's well-trained imagination, those mental skills are now being externalized into object-oriented software modules. The surgeon's delicate manual dexterity acquired through years of training is being coded into haptic interface modules that will accompany, guide, and in many cases assist the surgeon in carrying out a difficult procedure.

How will all this affect the heroic subject we've called "surgeon"? Will that new techno-supersurgeon be an upgrade on the last generation's heroic surgeon? The new surgeons would undoubtedly have background knowledge in the texts and practices of anatomy, biochemistry, physiology, and pathology, including some traditional practices from earlier generations. But they will require familiarity with, if not hands-on experience in, new fields such as biophysics, computer graphics and animation, biorobotics, and mechanical and biomedical engineering. They will also need to be aware of the importance of network services and bandwidth issues as enabling components of their practice. It is obviously unrealistic to assume that last generation's heroic surgeon will come repackaged with all these features, any more than next year's undergraduates will come to math class with slide rules. If we have learned anything about postmodern distributed production, it is to expect flat organizational structures, distributed teamwork, and modularization. Thus, given the complexity of all these fields, surgical systems will likely come packaged as turnkey systems. Many surgeons will be operators of these systems, performing "routine" cardiac bypass surgeries that implement predesigned surgical plans from a library of stored simulations owned by the company employing them. I don't mean that surgeons will simply become technicians or that surgery will cease to be a highly creative field. However, that creativity will be of a different sort, as many of the functions now internalized by surgeons are externalized into packaged surgical design tools just as computer-aided design packages such as Autocad, 3D Studio Max, or Maya have reconfigured the training, design practices, and creativity of architects. Some surgeons with access to resources will undoubtedly engage in high-level surgical design work, but that process will be mediated in teamwork involving software engineers, robotics experts, and a host of others.

Other specialties will be similarly altered by the medialization of surgery. Consider the impact on radiology. The radiologist has been crucial to the surgeon's ability to carry off a complex surgery prior to the age of medialization. Like the surgeon, the radiologist has been a highly valued and relatively autonomous agent. As a key professional in the surgical design process, the radiologist would make x-rays and more recently administer CT, MRI, and various other types of scanning modalities appropriate to the diagnosis of a suspected disease. Examining a dozen or so images, or a hundred or so slices of a CT or MRI scan, the radiologist would prepare a diagnostic report for the surgeon. Like the similar skill of the physician, the radiologist's diagnosis was heavily dependent on the keen observational skills required for detecting artifacts and spotting lesions or other abnormalities that would be the subject of the report. But the relative autonomy of the radiologist and his or her relationship to the diagnostic and surgical design process will certainly change in the near future.

As real-time computer-generated imaging becomes the norm, software tools for visualization and automated segmentation of tissues will displace the radiologist as interpreter of the data. Indeed, pressures are already mounting in this direction as the manufacturers of imaging systems such as GE, Siemens, and Brücke install systems that rapidly generate over a thousand images rather than a few dozen slices. Radiologists are currently under siege by an explosion of new data. Given the cardinal rule of data processing that valuable data should not go unused, the segmentation of this data into tissues, organs, and other anatomical structures, together with the detection of abnormalities, is becoming a problem for software automation. As automated tools for handling the explosion of imaging data arrive, the radiologist will undoubtedly reorient his or her professional activity and training to focus on new problems, such as the construction of surgical simulations. To do so, the radiologists will work closely with computer programmers and software engineers. Needless to say, if radiology as a medical specialty survives, the background, types of knowledge, and training of its practitioners will be radically different.

Should we deplore these developments? Many feel that the increase in technical mediation of surgery that I have discussed here, together with its attendant changes in organization, financing, careers, and personnel, are steps in the direction of the dehumanization of medicine by the advance of technology. To many, just describing these systems is in some sense to celebrate them, whereas our role as medical humanists should be to critique and wherever possible resist the technical interface driving a deeper wedge between caring doctors and their patients. While sympathetic to these views, I wonder where we might locate the moral high ground in order to fashion such a critical framework. The problem, as I see it, is that there is no "there" there to critique. The episodes I have treated illustrate that while the rapidity with which these changes are taking place may suggest creeping techno-logical determinism, this is anything but the case. Each of the technical steps I have described involves negotiations among a large network of actors, machines, and markets. The technology involved draws upon military-sponsored research in simulation, networking, and robotics. At the same time, it depends on imaging technologies driven by price reductions that derive from the entertainment industry, particularly improvements in three-dimensional computer graphics by leading-edge companies, such as Nvidia, who supply the video game industry.

The component technologies driving this surgical revolution are rapidly becoming ubiquitous. They are embedded in so many facets of our lives, from the tools of our workplaces, to our cell phones and personal digital assistants, to our means of entertainment, that it is impossible to identify the "good guys" and "bad guys." No less problematic are the values motivating the changes. Who can find fault with the professed goal of expanding the range of operable conditions, reducing blood loss and the danger of infection, and improving recovery times through advanced endoscopic procedures? Or, given the enormous costs of health care, who has a problem with the goal of making medical care efficient through training and simulation exercises linked to diagnostic and surgical procedures profiled to meet the needs of the specific individual? Haven't these goals always been the proper motivations of caring, humane medicine?

Perhaps more problematic for identifying a critical high ground is the phenomenon I have called "medialization." By this term I have sought to call attention to the ways in which the medical body is being redefined as the digital body. From stem cells to fully developed organisms, digital media provide the interface for medical intervention. But media are not transparent devices—and new media, with their increased involvement of all the senses, perhaps less so than previous media configurations. Media not only participate in creating objects of desire, they are desiring machines that shape us. Through medialization we come to desire the digital medical body. Media inscribe our situation: it is difficult to see how we can teleport ourselves to some morally neutral ground.

---

1   See J. Périssat, D. Collet, and R. Belliard, "Gallstones: Laparoscopic Treatment—Cholecystectomy and Lithotripsy: Our Own Technique," *Surgical Endoscopy* 4.1 (1990): 1–5, and F. Dubois, P. Icard, G. Bertholet, and H. Levard, "Coelioscopic Cholecystectomy: Preliminary Report of Thirty-six Cases," *Annals of Surgery* 211 (1990): 60–62.

2   For background on the PHANTOM system see the information at *http://www.sensable.com*.

3   For technical reports and news updates on the stages in development and approval of the Intuitive system see the archive section of the Intuitive Surgical, Inc., website: *http://www.intuitivesurgical.com*.

4   G. Herman and H. Liu, "Display of Three-Dimensional Information in Computed Tomography," *Journal of Computer Assisted Tomography* 1 (1977): 155–60.

5   M. W. Vannier, J. L. Marsh, and J. O. Warren, "Three-Dimensional Computer Graphics for Craniofacial Surgical Planning and Evaluation," *Computer Graphics* 17 (1983): 263–73.

6   M. W. Vannier, J. L. Marsh, and J. O. Warren, "Three-Dimensional CT Reconstruction Images for

Craniofacial Surgical Planning and Evaluation," *Radiology* 150 (1984): 179–84; J. L. Marsh, M. W. Vannier, and W. G. Stevens, "Computerized Imaging for Soft Tissue and Osseous Reconstruction in the Head and Neck," *Plastic Surgery Clinicians of North America* 12 (1985): 279-91; R. H. Knapp, M. W. Vannier, and J. L. Marsh, "Generation of Three-Dimensional Images from CT Scans: Technological Perspective," *Radiological Technology* 56.6 (1985): 391–98.

7   J. L. Marsh, M. W. Vannier, S. J. Bresina, and K. M. Hemmer, "Applications of Computer Graphics in Craniofacial Surgery," *Clinical Plastic Surgery* 13 (1986): 441–48; M. W. Vannier and G. C. Conroy, "Three-Dimensional Surface Reconstruction Software System for IBM Personal Computers," *Folia Primatologica (Basel)* 53.1-4 (1989): 22–32; M. W. Vannier, "PCs Invade Processing of Biomedical Images," *Diagnostic Imaging* 12.2 (1990): 139–47; M. W. Vannier and J. L. Marsh, "Craniofacial Imaging: Principles and Applications of Three-Dimensional Imaging," *Lippincott's Reviews: Radiology* 1.2 (1992): 193–209.

8   R. M. Koch, M. H. Gross, et al., "Simulating Facial Surgery Using Finite Element Models," *SIGGRAPH 96: Computer Graphics Proceedings*, Annual Conference Series (1996): 421–28.

9   C. A. Taylor et al., "Predictive Medicine: Computational Techniques in Therapeutic Decision-Making," *Computer Aided Surgery* 4.5 (1999): 231–47.

10   I. W. Hunter, T. D. Doukoglou, et al., "A Teleoperated Microsurgical Robot and Associated Virtual Environment for Eye Surgery," *Presence: Teleoperators and Virtual Environments* 2.4 (1993): 265–80.

11   Peter Oppenheimer and Suzanne Weghorst, "Immersive Surgical Robotic Interfaces," *Medicine Meets Virtual Reality (MMVR '99)*, San Francisco, CA, 1999.

12   Nathaniel I. Durlach and Anne S. Mavor, eds., *Virtual Reality: Scientific and Technological Challenges* (Washington, D.C.: National Academy Press, 1995), pp. 25–26.

13   Jerome P. Kassirer, "Managed Care and the Morality of the Marketplace," *The New England Journal of Medicine* 333.1 (July 6, 1995): 50-52, Thomas Bodenheimer, "The Oregon Health Plan — Lessons for the Nation, Part One," *The New England Journal of Medicine* 337.9 (August 28, 1997): 651–55, Thomas Bodenheimer, "The Oregon Health Plan—Lessons for the Nation, Part Two," *The New England Journal of Medicine* 337.10 (September 4, 1997): 720–23.

14   The classic sources on this point are Eliot Freidson, *The Profession of Medicine* (New York: Dodd, Mead, 1970); Magali Sarfatti Larson, *The Rise of Professionalism* (Berkeley: University of California Press, 1977); Charles Rosenberg, *The Care of Strangers* (New York: Basic Books, 1987); Paul Starr, *The Social Transformation of American Medicine: The Rise of a Sovereign Profession and the Making of a Vast Industry* (New York: Basic Books, 1982).

15   K. H. Höhne et al., *Voxel-Man 3D-Navigator: Inner Organs, Regional, Systemic, and Radiological Anatomy*, CD-ROM set (Berlin and Heidelberg: SpringerVerlag, 2000).

16   André Leroi-Gourhan and others have pointed out that a key feature in the construction of new media is the externalization of mental processes in an inscription device or system of inscription. See André Leroi-Gourhan, *Le geste et la parole. Dessins de l'auteur* (Paris: A. Michel, 1964). The relation of phonetic script to speech is the classical example of this phenomenon, but as Friedrich Kittler and others have pointed out, the process is evident in other inscription technologies. See Friedrich Kittler, *Discourse Networks 1800/1900* (Stanford, CA.: Stanford University Press, 1988) and Jacques Derrida, *Of Grammatology*, trans. Gayatri Spivak (Baltimore: Johns Hopkins University Press, 1976). For an excellent overview of the problem see David E. Wellbery's foreword to *Discourse Networks 1800/1900* (Stanford, CA: Stanford University Press, 1988).

The Irony of Virtual Flesh
Peter Oppenheimer

In his essay "The Virtual Surgeon: Operating on the Data in an Age of Medialization," Timothy Lenoir suggests that we will soon reach a time when the ubiquity of computers and the prevalence of virtual interfaces for surgical applications will lead to surgical practices where the "real" is indistinguishable from the "virtual." The provocative implication of Lenoir's essay goes beyond a "blending" of the real and virtual, however; it suggests that these two ontological states exist in a fruitful exchange that may lead to a fundamentally new space with new surgical practices, new professional guidelines, and new forms of fleshy existence. The virtual representation of human anatomy through digital media may well be primarily a stepping-stone to the cyber-mechanical production of artificial physical flesh. The loop from physical to virtual back to physical, thus closed, illuminates the fundamentally virtual nature of our physical bodies, if not all matter.

In order to understand how this space may be created, we need to understand how "data" and "flesh" interact in a mutually complementary system of exchanges: how moving from *in vivo* to *ex vivo* and back again through the use of silicon can actually lead to an *in vivo ex silica*.

In Vivo to *Ex Vivo*

Science fiction has depicted humanity's escape to a multisensory realm of experience as the holy grail of virtual reality research and development. These depictions have provided fertile ground for the fears and hopes of individuals who participate in a society saturated with information-processing technologies. Considering the long heritage and the prevalence of these projections, it may be best not to read them for the truth value of their claims but as sensitive indicators, navigating instruments really, for our exploration of technologically mediated spaces.

This conception of virtual existence as "mediated experience" has deep roots in the Western metaphysical tradition, and it is important to understand how virtual reality research has used this tradition to make sense of itself to itself. For instance, it is not coincidence that a virtual reality display system projecting on walls that surround the viewer is called a "cave." Obviously this refers to Plato's portrayal of the mediated quality of all sense experience in his allegory of the cave from *The Republic*.

By reanimating this reference we place current VR research as one more step in a long history that looks at mediated experience as the simulacrum, or the copy of the copy. And much in the same way that Plato did in the *Phaedrus*, we often frame this question in terms of the gains and losses of mediated versus bodily experience.

There is another tradition in the West that sees mediation itself as a new form of life. These ideas are cast in alchemical archetypes from Prometheus to Pygmalion, from *Frankenstein* to Genesis. This is a tradition that has explored the differences between representations and life and asks, what happens when life leaps from the page *ex vivo*? What happens when the pages of the book become the membranes for new forms of life?

Ex Vivo to *In Vivo*

Machines known as "replicators" can now generate solid physical objects from a computer-based three-dimensional numerical data representation. After processing the three-dimensional representation into stacks of two-dimensional contour maps, the machine renders the two-dimensional slices as layers of plastic. As the technology undergoes the evolutionary advances of the product cycle, solid three-dimensional objects will be manifested as readily as copies from today's laser printers.[1] Kuh-ching . . . kuh-ching . . . kuh-ching. . . . Artificial organs will be commodified if not grown.

Although related, this is very different from the dynamics of virtual experience outlined above. The three-dimensional numerical representation is not the endpoint of this production line but a soft blueprint of the final hardcopy. The map has now become the place. When this process is oriented toward the biological domain, the boundaries between natural and artificial flesh melt.

The Virtual Nature of All Objects

The blurring of the boundary between physical and virtual flesh reminds us of the virtual nature of actual physical objects. A mountain from our usual daily perspective is solid, solid, solid. But from the subatomic perspective it is mostly empty space. Visualization technologies allow us new conceptions of everyday objects. Which one is it, really? Is it solid or empty? To my mind it is both, or more precisely, solidness or emptiness is a quality of perspective rather than an inherent quality of the mountain. If we are conscious of what tools we are using to visualize objects, we can change our perspective when needed. We can thus learn that "everyday experience" is already virtual.

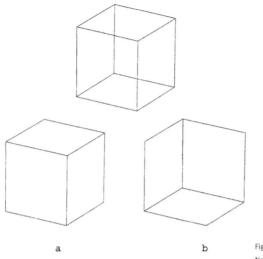

a   b   Figure 1.
                                                          Necker cube

A well-known demonstration of this is the Necker cube (fig. 1). A Necker cube has two possible three-dimensional interpretations but is just a set of lines drawn on a two-dimensional page. Which is it: three-dimensional interpretation a? Three-dimensional interpretation b? Or the lines on a two-dimensional page? The body is no different: it has a physical and a virtual interpretation. Although the brain does not so readily toggle between these interpretations as in the case of the Necker cube, the dualistic quality of our bodies remains.

Perhaps the best medical analogue to the "mostly empty space" interpretation of matter is the view of our bodies as a set of two-dimensional images stacked along an orthogonal axis. This technical, theoretically "objective" Cartesian representation is invasive to the coordinate systems of the functional structure of the body's organs. And yet this self-portrait of human anatomy is a beautiful neo-mythical expression of our nature, akin to the nuclear physicist's snapshots of scattering subatomic particles. We can see the space within the slices, but the detail of the slices may seduce us into losing track of what the slices constitute when put together. The inverse is also true; the beauty of the human form in total may seduce us, but we may thus lose track of the beauty of the spaces within the body.

In fact, the massive literature on metaphysical dualities (mind/body or otherwise) is also perhaps best viewed through the bifurcated lens of real virtual experience. In one sense, it forms a solid philosophical platform, but upon closer examination can also ultimately be perceived as the shaky ground of almost empty space.

In Vivo Ex Silica — New Virtual Space

What computer professional has not awakened in the morning to see the waking world boot up window by window? That structural representation of information confounds with a prior paradigm of things as solid objects for but a few dreamlike moments before solidifying into the Monday morning cup of coffee.

In much the same way, the virtual futures of medicine described by Lenoir will also solidify into new forms of space supporting existence for new forms of material entities. Can we orchestrate such blended paradigms to some advantage?

Virtual surgery will require new medical professionals possessing new skills. The radiologist, for instance, will be interpreting new kinds of imagery and will need a keener understanding of virtual reality technologies in order to discriminate whether an image feature of the now three-dimensional reconstruction is anatomically based or an artifact of the process.

Many of these skills will initially have to come from nontraditional settings. Can we successfully navigate these demands? Can we create fluid cultural environments that allow doctors to shift easily from viewing bodies as deconstructed slices to understanding the dynamics of how these slices fit together? Can a virtuoso video game player, pre-filtered for compassion and a sense of meticulous and fastidious thoroughness, be integrated into the surgical hyperteam whose players penetrate into the multidisciplinary and trans-hyper-phylum of posthuman cyborg? The medialization of medicine opens the possibility of such cross-referencing and repurposing of skill sets.

Transcending the Body's Duality

New materials will arise that will blend traditional metaphysical distinctions. New forms of semiotic flesh will be constituted through new forms of medical and material practice. If this can be done, then eventually the interpretation of body as mind can be reflected (à la Necker) into mind as body. "And in ultimate enlightenment, that duality fuses, vanishes, is transcended in Divine Ignorance."[2]

If one could not only embrace the dual nature of our bodies but also acknowledge the possibility of yet other unknown, if not unknowable, realities of our selves, then such a one may be an ultimate super doctor, founded in compassion and in an understanding of our multifaceted and ever mysterious nature.

---

1   Martin J. Moylan, "A Sci-fi Dream Gets Real," *Seattle Times*, July 26, 2001, p. C1.

2   Da Avabhasa, *The Heart's Shout* (Clearlake, CA: Dawn Horse Press, 1993), p. 158.

Flesh and Metal:
Reconfiguring the Mindbody in Virtual Environments

N. Katherine Hayles

In my recent book *How We Became Posthuman*, I struggled to avoid the Cartesian mind/ body split by making a distinction between the body and embodiment. But having made the analytical distinction between the body and embodiment, I could not escape the clay of dualistic thinking that clung to me regardless of how strenuously I tried to avoid it. In this essay, rather than beginning dualistically with body and embodiment, I propose instead to focus on the idea of relation and posit it as the dynamic flux from which both the body and embodiment emerge. In this view embodiment and the body are emergent phenomena arising from the dynamic flux that we try to understand analytically by parsing it into such concepts as biology and culture, evolution and technology. These categories always come after the fact, however, emerging from a flux too complex, interactive, and holistic to be grasped as a thing in itself. To signify this emergent quality of the body and embodiment, I will adopt the term proposed by Mark Hansen to denote a similar unity, the mindbody.[1]

My argument further implies that these co-evolutionary dynamics are not only abstract propositions grasped by the conscious mind but also emergent dynamic processes actualized through interactions with the environment. And here there is a problem. Especially in times of rapid technological innovation, there are many gaps and discontinuities between abstract concepts of the body, experiences of embodiment, and the dynamic interactions with the flux of which these are enculturated expressions. The environment changes and the flux shifts in correlated systemic and organized ways, but it takes time, thought, and experience for these changes to be registered in the mindbody. Bridging these gaps and connecting these discontinuities is the task taken on by the three virtual reality artworks discussed here: *Traces* by Simon Penny and his collaborators, *Einstein's Brain* by Alan Dunning, Paul Woodrow, and their collaborators, and *NØtime* by Victoria Vesna and her collaborators.

If art not only teaches us to understand our experiences in new ways but actually changes experience itself, these artworks engage us in ways that make vividly real the emergence of ideas of the body and experiences of embodiment from our interactions with increasingly information-rich environments. They teach us what it means to be posthuman in the best sense, in which the mindbody is experienced as an emergent phenomenon created in dynamic interaction with the ungraspable flux from which also emerge the cognitive agents we call intelligent machines. Central to all three artworks is the commitment to understanding the body and embodiment in relational terms, as processes emerging from complex recursive interactions rather than as pre-existing entities. Because relationality can be seen through many lenses, I have chosen works that place the emphasis on different modes of relation. *Traces* foregrounds the relation of mindbody to the immediate surroundings by focusing on robust movement in a three-dimensional environment; *Einstein's Brain* foregrounds perception as the relation between mindbody and world that brings the flux into existence for us as a lived reality; and *NØtime* emphasizes relationality as cultural construction.

These configurations can also be understood in terms of the typology Don Ihde proposes in *Technology and the Lifeworld: From Garden to Earth*.[2] Parsing the general situation as Human-technology-World Relations, he identifies three variants. The first, which he calls "embodiment relations," bundles the human and the technological into one component and emphasizes the relationality between this component and the world: (Human-technology) → World. This corresponds to the situation in *Traces*, in which the technology re-produces the human form in a simulation (thereby bundling together the technology and the human) and emphasizes the movement of the simulated techno-human through space. The second variant, identified by Ihde as "hermeneutic relations," bundles together technology and the world and emphasizes the relationality with the human: Human → (technology-World). This parsing is performed in *NØtime*, which leaves the human body unencumbered to experience an installation space permeated by sensors, actuators, and display technologies, which bundle together the world and technology. The third parsing, "alterity relations," is parsed as Human → technology-(-World), where the parentheses around World indicate that the bundling of technology and world is achieved through the creation of a simulated world, a configuration enacted by *Einstein's Brain*. Although the three emphases of enactment, perception, and enculturation (corresponding to embodiment, hermeneutic, and alterity relations in Ihde's schema) by no means exhaust the ways in which relationality brings the mindbody and the world into the realm of human experience, they are capacious enough in their differences to convey a sense of what is at stake in shifting the focus from entity to relation.

Relation as Enactment

In *The Embodied Mind: Cognitive Science and Human Experience*, Varela, Thompson, and Rosch articulate a vision of relationality that has much in common with *Traces*. They write, "living beings and their environments stand in relation to each other through *"mutual specification* or *codetermination."*[3] Coining the term "enaction" to describe this dynamic interplay between self and world, they envision mind-body and environment coming into existence through a mutual process of "codependent arising" (110).

It is precisely this kind of relationality that Simon Penny wanted to implement in a virtual reality (VR) environment. As early as 1994 he articulated a desire to depart from the usual VR model that in his assessment "blithely reifies a mind/body split that is essentially patriarchal and a paradigm of viewing that is phallic, colonizing, and panoptic."[4] In *Traces*, Penny along with collaborators Jeffrey Smith, Phoebe Sengers, André Bernhardt, and Jamieson Schulte created an interactive artwork designed to bring the body more fully into the virtual space. Reacting against the VR rhetoric of disembodiment, they critique this rhetoric as deriving from "an essentially uninterrogated Cartesian value system, which privileges the abstract and disembodied over the embodied and concrete."[5] They propose, by contrast, to build "an unencumbering sensing system which modeled the entire body of the user" (3).

Working with a three-dimensional CAVE environment that displays simulated visual images along four surfaces (three walls and the floor) as well as in the goggles of the user, they implemented a visual tracking system that computes the volume of the user's body by modeling its movement in space and time through three-dimensional cubes called voxels (volumetric units named by analogy to two-dimensional pixels). From this computation they created "traces," simulated images of volumetric residues that trail behind the rendered model of the user's body, gradually fading through time as continued movement creates new traces that also fade in turn. The body model and residues are composed of lilac-colored voxels five cm on a side, rendered in a simulation space 60 x 60 x 45 voxels in three dimensions, with computational time steps of fifteen frames per second (fps). Since the time step interval falls below the threshold of twenty-four fps at which frames appear to human viewers as continuous motion, the computational process manifests itself to the user and outside viewers as somewhat jerky in its motion. This effect, although in one sense a constraint imposed by the amount of computation required for each updating, is embraced by Penny and his collaborators as part of a larger aesthetic strategy to "avoid any pretense of organic form. There was a desire to be up-front about the fact that this was a computational environment, not some cinematic or hallucinatory pastoral scene" (13). For the same reason, the collaborative team renounced what it regarded as the "eye candy" of virtual worlds ready for exploration and used texture-mapping only for the virtual room projected on the CAVE walls, about twice the size of the CAVE's dimensions of three meters on a side. Instead of a graphically rendered virtual world, the user "simply sees graphical entities spawned by various parts of their [*sic*] body when in motion" (10).

The avatar interface is designed to be, in Penny's terminology, "autopedagogical," teaching the user how to interact with it through a three-phase evolution: Passive Trace, Active Trace, and Behaving Trace. In the Passive Trace the simulated volume passively follows the user's motions, creating the impression that the user "dances" a "sculpture" (4), though a transitory one that gradually fades into transparency as it moves away from the viewer into the simulated space on the front wall of the CAVE. With its time-sensitive evolving transparency, the Passive Trace would seem to have more in common with Bergson's vision of flowing time than with the enduring static form that sculpture usually implies. As the user continues to move in the space, the trace transforms from a passively following cloud into active entities that can be "spun off" the user's body through rapid motion or acceleration, for example by flicking one's hand rapidly down one's arm as if shaking off water droplets. At first these entities follow the user, but gradually they become more autonomous as their motions are guided by autonomous agent software. As they make the transition into the Behaving Trace, they exhibit behaviors characteristic of such artificial life programs as Craig Reynolds's "Boids," simulated forms that exhibit flocking behavior when programmed with a relatively simple set of rules such as "always fly toward the center of where the other objects are." The simulated objects in the Behaving Trace may follow the user, for example, or they may break off and head in other directions, moving as a flock following its own artificial life dynamics.

Although the immediate meaning of "autopedagogical" for Penny is the progression whereby *Traces* teaches the user how to interact with it, the term evocatively points toward other realizations as well. By incorporating a temporal dimension into the work, and especially by having the duration of the trace visibly fade away as it ages, the artwork resists the fantasy that information technologies will allow us to escape our bodies and move into transcendent spaces where we can escape the ravages of time. Another realization emerges from the "traces" metaphor, which suggests new kinds of possibilities for interactions between humans and intelligent machines. As Penny and his collaborators point out, the *Traces* simulation, considered as an avatar, occupies a middle ground between avatars that mirror the user's motions and autonomous agents that behave independently of their human interlocutors. As the "trace" avatar transforms from mirroring the user's actions to engaging in autonomous behaviors, it enacts a borderland where the boundaries of the self diffuse into the immediate environment and then differentiate into independent agents. This performance, registered by the user visually and also kinesthetically as she moves energetically within the space to generate the entities of the Active and Behaving Traces, makes vividly clear that the simulated entities she calls "her body" and the "trace" are emergent phenomena arising from their dynamic and creative interactions.

Moreover, the elegant simplicity of the simulation—the refusal to add "eye candy" to the visual effects—helps to make real to the user that the avatar is in effect indistinguishable from the user's interface with the computer. The trace avatar, Penny and his collaborators write, "must be thought of as the part of the system which is intimately connected to the user. In this way, the line between system, avatar, and interface also becomes blurred; the avatar becomes the interface, the point at which the computational system and the user make contact" (22). The aesthetic strategy of refusing to conceal the computational nature of the simulation resists the fantasy Donna Haraway calls "transcendent second-birthing" by grounding the work in the constraints of real-life computational and sensing devices. As the *Traces* essay explaining the construction of the work makes clear, this is no illusory world of limitless possibilities but a carefully engineered artwork in which numerous trade-offs and "workarounds" must be made for the project to be feasible. The ingenuity, creativity, and skill of the designers and programmers are repeatedly tested as they come up against a variety of problems, from devising a workable camera tracking system to balancing the graininess of the voxelated image against increased latency times as the computational load increases. Coming to grips with these problems, they achieve the key insight that constraints can function as opportunities as well as problems. This approach led them to see that the problems involved in having the avatar exactly track the user's body could become constructive with a change of metaphors. Rather than regarding the avatar as a mirroring puppet, they think of it as a trace emerging from the borderlands created by the energetic body in motion. What was a tracking problem is thus transformed into the possibility of creative play between user and avatar.

The net result of these compromises, creative solutions, and transformations is to make real for us the insight that the artwork is not simply the instantiation of an abstract concept but an artifact that emerged from the dynamic relation between the vision of the designers, the constraints imposed by the situation, and the powerful but still limited capabilities of the intelligent machines that perform the sensing, computational, and rendering tasks that make the project a reality. In its form, construction, and functionality, *Traces* testifies to this relationality at the same time that it also performs relationality for the user. Far from the fantasy of disembodied information and transcendent immortality, *Traces* bespeaks the playful and creative possibilities of a body with fuzzy boundaries, experiences of embodiment that transform and evolve through time, connections to intelligent machines that enact the human-machine boundary as mutual emergence, and the joy that comes when we realize we are not isolated from the flux but rather enact our mindbodies through our deep and continuous communion with it.

Relation as Perception

In "The Brightness Confound," Brian Massumi reminds us of Wittgenstein's anecdote about his puzzlement as he stares at his sunlit table and realizes that he cannot say what color the table is, for his perception fuses luminosity, radiance, and color into a unity that defies tidy categorization. Following Marc Bornstein, Massumi calls this unity "the brightness confound" and makes the simple but elegant point that the perception is a "singular confound of what are described empirically as separate dimensions of vision."[6] In this sense the confound is absolute.

> Absoluteness is an attribute to any and all elements of a relational whole. Except, as absolute, they are not "elements." They are parts or elements before they fuse into the relational whole by entering indissociably into each other's company; and they are parts or elements afterwards, if they are dissociated or extracted from their congregation by a follow-up operation dedicated to that purpose. In the seeing, they are absolute. (82)

Insisting on this absoluteness, Massumi nevertheless writes as if the elements of the confound have a prior existence as separate entities. Humberto Maturana expunges this vestige of realism when, in developing the theory of autopoiesis, he makes the point that someone experiencing a hallucination would be unable to distinguish the hallucination from reality.[7] In Maturana's view, what we perceive is reality for us.

In the *Einstein's Brain* project, Canadian artists Alan Dunning and Paul Woodrow stage what we might think of as Maturana's claim (although they come to this view via their own independent paths and not necessarily through Maturana). They are keenly conscious of the ironic overtones of their chosen title, referencing a similarly titled essay by Roland Barthes. Commenting on Einstein's brain as a fetishized object, they write, "His brain has passed into the world of myth, cut up and minutely examined but revealing little."[8] The title points up the fact that the brain as fetishized physical object, considered in isolation from the world, cannot possibly account for the richness of human experience. In his meditation on the subject, Barthes related the duality of physical brain and prodigious mind to a split between Einstein as the researcher and Einstein the knower of the world's innermost secrets.[9] Rooted in the physical brain, Einstein's mind nevertheless seemed to have nearly occult powers of insight, at least in the popular imagination. This oscillation between ordinary physical reality and occult power translates in the *Einstein's Brain* project into a desire to use advanced technology to reveal the constructedness of our everyday world.

The *Einstein's Brain* project has been in process for five years and has taken a variety of forms in different installations, but a common idea unites all the instantiations.[10] The artists (like Maturana) are committed to the realization that the world of consensual reality does not in any sense exist "out there" in the forms in which we perceive it. Rather, the world we know is an active and dynamic construction that emerges from our interactions with the flux.

> We think of the body as separate from the world—our skin as the limit of ourselves. This is the ego boundary—the point at which here is not there. Yet, the body is pierced with myriad openings. Each opening admits the world—stardust gathers in our lungs, gases exchange, viruses move through our blood vessels. We are continually linked to the world and other bodies by these strings of matter. We project our bodies into the world—we speak, we breathe, we write, we leave a trail of cells and absorb the trails of others. The body enfolds the world and the world enfolds the body—the notion of the skin as the boundary to the body falls apart.[11]

They self-consciously position their work in opposition to military and corporate uses of virtual reality, which continually aim for greater and greater realism. They point out that when virtual reality illusions are engineered with the goal of seamlessly reproducing the "real" world, the effect (wittingly or not) is to reinforce existing structures of authority and domination—structures that in their desire to preserve the status quo find it in their interest to foreclose alternative constructions of reality and moreover to keep them from coming to mind as possibilities. By contrast,

Dunning and Woodrow conceive of the simulation technologies as deliberately imperfect, so as to make clear their construction as "reality engines connected not to the generation of a reality but as a means of attending to a consciousness that in turn fashions a reality" (7).

To resist the domination in VR of a "realism rid of expression, symbol or metaphor . . . sustained by the authorities of homogeneity and seamlessness" (1), Dunning and Woodrow create a "cranial landscape" merging symbolic and semiotic markers with the landscape of experience (5). Their work often has a somewhat idiosyncratic range of reference overlaying the consistency of their vision, rather as if a magpie had collected shiny bits from here and there that attracted her attention and then wove them into a nest of breathtaking coherence and careful design. So the inspiration for the cranial landscape comes partly from the *Carte du tendre*, Madeleine de Scudèry's 1654 romantic narrative map inscribing onto a landscape names indicating the predictable heating up and cooling off of a love affair. Appropriating the name of one of these sites, the Forest of Vowels, Dunning and Woodrow create a semiotically marked landscape that exists for the user as a negotiable surface and also as a changing landmass tied in with the user's responses as they are registered through a reading of her brain waves and other physiological indicators.

Another source of inspiration is the *dérive* of the Situationists, a random walk through the city governed by the principle that every turn and meander should be taken at random rather than with the intent to arrive somewhere. Guy Debord's 1957 map of Paris showing arrows marking the course of a *dérive* suggested to Dunning and Woodrow that even something as apparently static and durable as city architecture might be re-imagined as emerging in complex interplay with human enactments. So their plan for *Forest of Vowels* calls for the association of external events in the real world with the landscape of the virtual world. These associations include feeding in the moon's changing gravitational forces so they alter the form of the virtual world's landmasses, tying fluctuations in the stock market to the growth patterns of trees and plants, and connecting the daily attendance figures at Graceland to the changing cultural paradigms of the virtual world (6).

Another idiosyncratic influence is *The Stone Tapes*, a story produced by Nigel Kneale on BBC Television on December 25, 1972. The story takes the form of a mystery centered on an apparently haunted building, which has somehow recorded in its inorganic stone traumatic events that happened there; the building has the capacity to play these recordings back by transferring them directly to the brains of people who come inside the building, so that it appears to them as if they are spectators of the original events. The story appeals to Dunning and Woodrow on multiple levels. The building displays qualities that make it appear as if it can operate as a subject, thus blurring the boundary between an exterior static architecture and a dynamic interior world of human emotion. Moreover, the humans haunted by the building must confront the apparent reality of an illusion generated from inside their own brains, blurring the boundary between their perceptions of the real world and the illusions activated for them by the stone building.

The incongruities between a virtual reality experienced by those who are "haunted" and the consensual reality experienced by on-lookers are dramatically staged in Dunning and Woodrow's installation *The Madhouse*. In this installation, participants wearing VR goggles engage in behaviors that can only appear strange and bizarre from the viewpoint of those who do not see or hear the simulations. In a similar mode is their plan for an artwork which, by creating deliberately unstable and deficient renderings of the virtual world, would force participants to question consensual reality. In this projected work, vision is blurred, detail is shifting and inconstant, slower or faster frame rates suggest a rendering engine behind the scenes, left- or right-hand sides of stereoscopic vision blink out, depth perception is lost, objects only appear when the viewer is in motion, the edges of the worlds visibly reinvent themselves. They write of their motivation for these kinds of strategies:

> As western artists, we developed from a world where we learned to objectify our bodies, to separate our minds from our bodies and viscera, where we learned to distinguish matter from mind and where the construction and placement of objects was the focus and culmination of our intentions and desires. Developments in cultural and social theory and in technology have suggested that we and other artists shift their attention away from a graspable, predominately corporeal world to one which is increasingly slippery, elusive and immaterial. Mind and matter, combining in the cognitive body, are interdependent. The world we inhabit is in flux, comprised of increasingly complex connections and interactions. In this world there are no fixed objects, no unchanging contexts. There are only coexistent, nested multiplicities. (8)

In the "unbroken field of transformations" that for them constitutes the emergent dynamic we call reality, virtual reality art can play a vital role in shaking the belief that our bodies and the world exist independent of relation. They intend their art to enact engagements that make vividly real the fact that everything in our world, including (or rather especially) the human mindbody, emerges precisely from our relation with the ongoing flux.

These ideas and strategies come together in the installation due to open at the TechnOboro Gallery in Montreal in September 2001 and exhibited in prototype at the Digital Arts Conference at Brown University in April 2001, where I saw it. The centerpiece of the installation is the ALIBI, the "anatomically lifelike interactive biological interface," an anatomically correct life-size model of the human body stuffed with a wide variety of sensors, including theremin proximity sensors, touch sensors, aroma sniffers, pressure sensors, sound sensors, and carbon dioxide sensors. Participants wear goggles that can be arranged to show only the simulated world or (by removing the blinders from the lens area) convert the scene to a "mixed reality" in which both the simulation and the real room are visible. They are able to see simultaneously the virtual reality projection and the *artifactual* body, which lies on a light table in the center of a room. Made from a cast of a male model, the body is painted with thermochromic paint that appears as a lovely dark blue when cool but turns white when warmed by someone's touch, fading again to blue as that area cools to the ambient temperature.

Participants can interact with the body by touching it on the thigh, abdomen, legs, etc., by whispering in its ear, even breathing into its mouth. These interactions, when sensed by the system, activate and change the simulated worlds being imaged in the goggles. The blue body thus acts as a navigational interface, opening portals to a variety of simulated worlds as the appropriate body areas are touched, massaged, and otherwise manipulated. One user wears a helmet capable of sensing her electroencephalic activity, including alpha, beta, theta, and delta brain waves. These data are fed into the simulation, along with other biological data collected from the user such as blood pressure, pulse rate, and galvanic skin response. The data trigger the performance of simulated images, with sunbursts, polygons, and flashes of light appearing in response to the user's reactions. Moreover, the amplitude and frequency of the participant's brain waves are converted to MIDI files and used to create a soundscape for the simulation, which serves as an acoustic transform of her ongoing physiological responses.

Two other components complete this complex work. On the back wall are projected images of revolutionary historical events, including authentic footage of statues being pulled down during the Russian Revolution and speeches by Mao. This corresponds to the *Stone Tapes* motif, establishing a visual connection between past and present and "haunting" participants with events that have changed the course of history and that continue to be remembered as dramatic instantiations of the revolutionary impulse. Further complexity is added by the presence of a "viewing room," in which other visitors can see the artifactual body being manipulated by users who engage in actions and behaviors that remain inexplicable to those who cannot see the virtual worlds. The effect, once again, is to call into question consensual reality by fragmenting the space so that different versions of "reality," virtual and actual, compete and conflict in their representational stimuli.

The most striking aspects of the installation from my point of view are the feedback loops between the user's responses, her interactions with the artifactual body, and the production of the simulated world. Imagine the scene. You are in some initial brain state that generates images and sounds in the simulated world you see. While watching these displays, you begin to touch the body in its sensitive areas, opening portals to other simulated worlds, which trigger new responses in you, which feed back into the simulation to change it, which makes you want to touch the body in new ways, which further changes the simulated images and sounds, which in turn generate yet more responses from you. The loop is endless, and endlessly fascinating, forming, as the authors say, a "single intelligent symbiotic system." [12]

Commenting on a different artwork, Dunning and Woodrow explain the relation of the body to the VR world in ways that apply with special force to this installation.

> These worlds are not external to the body, but, are properly thought of as being inside the body. This accounts for the apparent invisibility of the body in a virtual space. The body disappears because it is turned in on itself. The ego-boundary is no longer the point at which the body begins and ends in relation to an external environment, but is, rather, . . . the very limit of the world. [13]

Using a different metaphor in "The Stone Tapes, the Derive, the Madhouse," they remark, "It is as if we are inside ourselves, like a three-dimensional eye which constructs itself as it moves through internal haptic space." [14] Relationality here is not beside the point; it is the point of a mindbody that realizes itself through its playful and intense interactions with evolving virtual worlds, which in the view of these artists include our perceptions of the real world as well as our experiences of simulated ones. In this sense all human experience is a "mixed reality," emerging from another kind of brightness confound in which technology, the world, and human embodiment all play a role.

Relation as Enculturation

A notable characteristic of *NØtime* is its collaborative nature, a feature of *Traces* and *Einstein's Brain* as well. The human collaborators, credited by name and sometimes by affiliation, indicate the range of expertise necessary to construct the installations and include such skilled contributors as computer scientist and cultural critic Phoebe Sengers for the *Traces* project, software designer and computer science engineer Hideaki Kuzuoka for the *Einstein's Brain* project, and software designer Gerald Jong for the *NØtime* project. For convenience I have referred to works using the names of the artists who had the initial idea as creator, but they more than anyone else realize how deeply their collaborators have shaped the projects and how essential their contributions have been. Less prominently featured and usually identified by model name and technical capacity are the silicon collaborators, the intelligent machines and software packages without which these works would have been impossible to create. It is not merely whimsical to refer to the machines as collaborators, for their capabilities and limitations are as important to the project's shape as the capabilities and limitations of the human designers. These silicon collaborators include computers, software programs, sensing systems, music synthesizers, tracking systems, motion detectors, and a host of other processors, interfaces, and actuators.

The complexity of the collaborations between many different humans and many different intelligent machines indicates that in a deep sense all of these projects are distributed cognitive systems. Moreover, cognition takes place not in the minds of the artists and the logic gates of the machines but also in the participants who interact with the artworks. As *Einstein's Brain* in particular makes clear, the user's interactions with the installation are not merely passive viewing of pre-existing works but active components in the work's construction. Among these three artworks, *NØtime* insists most visibly and interactively on the distributed cognitive collaborations that construct it and especially on the role played by the global community of intelligent machines we call the Internet. It also locates the arena of relationality with which it is concerned in the broadest geographic terms. Whereas *Traces* focused on the immediate proximity of the body and *Einstein's Brain* on the room-sized spaces where the artifacts were placed and the simulations projected, the reach of *NØtime*'s enactments is global, although it simultaneously insists on the importance of local interactions and proximity.

Victoria Vesna originally conceived of *NØtime* as a response to the common postmodern condition of having no time, of living our lives amid the multiple conflicting commitments and stresses that people negotiating complex urban environments find to be an inevitability of contemporary life. Her playfully paradoxical idea was to create avatars that could take over portions of our lives and live them for us while we were busy doing other things. As the project evolved, the idea of collaborative interactions that together create a "person" or a "life" took a somewhat different turn, focusing on a nested series of relations between the local and the remote, the individual and the collective, the proximate and the distributed, the immediate and the long-term. As with *Traces* and *Einstein's Brain*, the effect is to create a space of intense interaction and feedback in which the subject experiences herself as emerging from relational dynamics rather than existing as a pre-given and static self.

The artwork consists of a distributed cognitive system that includes a physical installation located in a gallery space and a remote component played out over the Internet. The physical installation consists of a gridwork of five supporting legs and four trusses that, draped with cream-colored spandex fabric, forms a beautifully translucent three-dimensional spiral thirteen feet high, which a visitor enters with the dawning delight of a snail discovering a shell of palatial dimensions. When the visitor positions herself at the center of the installation, the translucent sheeting functions like a borderland between inside and outside, for it creates a sense of enclosure at the same time as it allows shapes and sounds to be discernible through the fabric.[15] On the wall is a projection flashing the names of participants who have previously created "bodies" in *NØtime*. When the visitor sees a name she recognizes or likes, she steps forward and the "body" corresponding to the name is displayed on another wall.

Like Penny, Dunning, and Woodrow, Vesna is critical of the tendency in military and corporate VR to move toward greater and greater realism. Rather than participate in this tendency by creating an anthropomorphic avatar, Vesna prefers to break with realistic representation and visualize the information/energetic "body" as a tetrahedron consisting initially of the six lines and four apexes required to outline the tetrahedral shape. The tetrahedron, messages flashing on the walls explain, is so privileged because it alone of all polygons has the greatest resistance to an applied load. When the load exceeds the critical tolerance, a tetrahedron will not dimple or bend like other polyhedral structures. Rather the tetrahedron turns inside out, thus making it "unique in being its own dual." These characteristics led Buckminster Fuller to choose the tetrahedron as the basic unit of construction for geodesic domes. They also relate to the tetrahedral shape of carbon stereochemistry, which makes the tetrahedron the essential shape for all carbon-based life on earth. Vesna calls the six edges of the tetrahedron "intervals" and associates them with the components essential to human life as identified by the Indian chakra system, adding color coding so that the family interval appears red, the finances interval orange, the creativity interval yellow, the love interval green, the communication interval blue, and the spirituality interval purple.

The apexes are also named, but here the naming scheme focuses on the cultural constructions that Richard Dawkins called "memes," that is, ideas, jingles, and images (think of the smiley face) that propagate rapidly through the culture, acting as ideational viruses that use humans as their conceptual replication system, much as Dawkins envisioned the "selfish gene" as doing through the physical body.[16] When a user creates a body as tetrahedron, she chooses the length of the intervals, a choice that reflects the relative importance she gives to the six components and determines the shape of that particular tetrahedron. In addition, this initial choice affects the overall shape as other tetrahedrons are added onto the virtual "body." To complete the body, the user names the four apexes with words or short phrases representing the memes she wants to circulate through virtual space by means of her "body." Once the body is complete, it is correlated with a three-dimensional soundscape that the on-site visitor can navigate by changing position within the installation. Gerald Jong's custom software, entitled "fluidiom," coordinates the visitor's position (as registered by motion sensors) with this soundscape, creating an acoustic experience unique to the interactions between a specific virtual body and a user's unique movements within the space.

Once the basic body is constructed, it grows through collaborators who are willing to spend time in the physical space. The longer an on-site visitor stays gazing at someone's tetrahedron, the more intervals are added to the figure. The body's owner can then add more memes, allow friends to add them, or distribute cards that enable visitors to add them at an on-site Internet connection. In keeping with the installation's theme, however, growth cannot continue indefinitely. Enacting the finitude that makes time, space, and life span limited commodities for all humans, a body deconstructs when it reaches a size of 150 intervals. The event is announced in advance at the web site, and people are invited to witness the virtual collapse. At that point the overgrown body is archived in a file accessible only to the owner, who has the option to start the growth process again with the same basic tetrahedron or to craft another one.

Through its distributed architecture, collaborative procedures, and sculpturally striking on-site installation, *NØtime* enacts the human body as an emergent phenomenon coming into existence through multiple agencies, including the owner's desires, the cultural formations within which identities can be enacted and performed, and the social interactions that circulate through the global networks of the World Wide Web. The phenomenon of "no time" is thus transformed from a generator of stress and an indicator of a declining quality of life into a site for creative play and collaborative interaction. But only, of course, if we make the time to visit the installation, participate in the web site, and extend the bodies of our fellow humans by physically committing ourselves to relational interactions that last longer than the thirty seconds usually accorded a gallery installation. Relationality requires care, attention, and dynamic interaction, all of which take the time that *NØtime* paradoxically insists we have after all.

Relation as the Posthuman

In *How We Became Posthuman*, I argued that developments in such fields as cognitive science, artificial life, evolutionary psychology, and robotics were bringing about an understanding of what it means to be human that differs so significantly from the liberal humanist subject that it could appropriately be called posthuman. Among the qualities of the liberal humanist subject displaced by technoscientific articulations of the posthuman are autonomy, free will, rationality, individual agency, and the identification of consciousness as the seat of identity. The posthuman, whether understood as a biological organism or a cyborg seamlessly joined with intelligent machines, is seen as a construction that participates in distributed cognition dispersed throughout the body and the environment. Agency still exists, but for the posthuman it too becomes a distributed function. Consciousness for the posthuman ceases to be seen as the seat of identity and becomes instead an epiphenomenon, a late evolutionary add-on whose principal function is to narrate just-so stories that often have little to do with what is actually happening.

In the crises precipitated by the deconstruction of the liberal humanist subject, one kind of response is represented by attempts to reinstate the lost qualities through mastery of increasingly powerful computational and informational technologies. If consciousness is reduced to an epiphenomenon, perhaps its sovereign role can be reinstated by losing the body and resituating the mind within a computer. If agency, like cognition, is distributed, perhaps it can be regained by creating more powerful prostheses, more extensive implants, more smart weapons. These responses share a reluctance to accept human finitude; they remain intent on imposing the will of the individual onto the world seen as an object to dominate. In these constructions, the subject remains inviolate even while losing the body, and the boundaries of the subject continue to be clearly delineated from an objective world. In an important sense, these responses carry on the worst aspects of the liberal humanist subject even as they turn toward the posthuman.

Another kind of response is enacted by the virtual reality artworks discussed above. Here the posthuman is embraced as the occasion to rethink the mind/body split and the premise that mind and body, like the rest of the world, pre-exist our experiences of them. As we have seen, the relational stance enacted by these works puts the emphasis instead on dynamic interactive processes from which both mindbody and world emerge together. The significance of these works in this posthuman moment is profound, for they operate with a performative intensity that makes us realize the importance of emergent relationality in mind and body, transforming these two "elements" into the mindbody that in turn is embedded in our relations with the techno-world. Speaking to more than conscious mind, these artworks provide our mindbodies with rich experiential fields that invest the relational stance with meanings that work on multiple levels, including the neocortex but reaching below and beyond it as well. They vividly demonstrate the promise of the posthuman, which despite all its problems and dangers may open us to the realization that without relation, existence (if it is conceivable at all) would be a mean and miserable thing. We do not exist in order to relate; rather, we relate in order that we may exist as fully realized human beings.

I am grateful to Michael Fadden and Carol Wald for help with the research for this article, and to Simon Penny, Alan Dunning and Paul Woodrow, and Victoria Vesna for sharing with me unpublished essays, videos, pre-exhibit showings and other materials concerning their artworks.

1   Mark Hansen, presentation at University of California at Los Angeles, May 2001.

2   Don Ihde, *Technology and the Lifeworld: From Garden to Earth* (Bloomington: Indiana University Press, 1990), p. 106 and passim.

3   Francisco Varela, Evan Thompson, and Eleanor Rosch, *The Embodied Mind: Cognitive Science and Human Experience* (Cambridge: MIT Press, 1991), p. 198.

4   Simon Penny, "Virtual Reality as the Completion of the Enlightenment Project," *Culture on the Brink: Ideologies of Technology*, edited by Gretchen Bender and Timothy Druckrey (Seattle: Bay Press, 1994), pp. 231–63, especially p. 238.

5   Simon Penny, Jeffrey Smith, Phoebe Sengers, André Bernhardt, and Jamieson Schulte, "Traces: Embodied Immersive Interaction with Semi-Autonomous Avatars," unpublished essay, p. 2. I am grateful to Simon Penny for giving me permission to quote from this essay prior to its publication.

6   Brian Massumi, "The Brightness Confound," in *Body Mécanique: Artistic Explorations of Digital Realms*, ed. Sarah J. Rogers (Columbus: Wexner Center for the Arts, 1998), pp. 81–94, especially p. 81.

7   In "Biology of Language: The Epistemology of Reality," in *Psychology and Biology of Language*, ed. George A. Miller and Elizabeth Lennenberg (New York: Academic Press, 1978), pp. 27–63, Humberto R. Maturana writes, "for the operation of the nervous system (and organism) there cannot be a distinction between illusions, hallucinations, and perceptions" (p. 46). In my view this claim needs to be modified. Some people who experience migraine headaches see visual auras but are quite conscious these auras are not reality. Other visions, such as those inscribed by Hildegard in *Scivias*, are taken by the perceiver as real but are distinguished from ordinary reality, in this case by being identified with the divine. A similar case is presented by the auditory hallucinations experienced by the science fiction writer Philip K. Dick near the end of his life, which he finally decided were communications from an extraterrestrial intelligence. Oliver Sacks in *The Man Who Mistook His Wife for a Hat* (New York: Harper Perennial, 1990) reports auditory and visual hallucinations that seemed very real to the patient but were also understood by the patient as coming from somewhere other than consensual reality, in one case as replayed memories (pp. 132–55).

8   Alan Dunning and Paul Woodrow, "Einstein's Brain," at *www.acs.ucalgary.ca/~einbrain/EBessay.htm*, p. 2.

9   Roland Barthes, "The Brain of Einstein," *Mythologies* (New York: Hill and Wang, 1971), pp. 68–70.

10   Collaborators vary from project to project but include, among others, Martin Raff of the MRC Laboratory for Cell Biology, University College, London; Pauline van Mourik Broekman, *Mute* magazine, London; Hideaki Kuzuoka, Department of Engineering, University of Tsukuba, Japan; Nick Dalton, Bartlett School of Architecture, University College, London; and Arthur Clark, Department of Neurology Health Sciences, University of Calgary, Calgary AB Canada.

11   Dunning and Woodrow, "Einstein's Brain," p. 7.

12   Alan Dunning and Paul Woodrow, "The Stone Tapes, the Derive, the Madhouse," presented at the New Media Institute at the Banff Centre, September 2000, p. 6.

13   Dunning and Woodrow, "Einstein's Brain," p. 5.

14   Dunning and Woodrow, "Stone Tapes," p. 3.

15   Don Ihde in *Technology and the Lifeworld*, p. 72 ff., remarks that many people want the advantages of technology without having it intrude on their lives, a contradictory desire made manifest in the wish that powerful technologies exist but that they also be transparent. The translucent enclosure seems to acknowledge this wish and also resist it by evoking transparency and simultaneously denying it.

16   Richard Dawkins, *The Selfish Gene* (New York: Oxford University Press, 1990).

Distributed Systems: Of Cognition, of the Emotions

Kathleen Woodward

In her excellent essay, Hayles discusses three virtual reality artworks that invite us to experience emerging structures of culture in our information-saturated environments. Her focus is on new concepts of the body and new experiences of embodiment, and her intent is to explicitly reject the analytical distinction between the mind and the body and to stress the concept of relation between the environment and what she calls, following Mark Hansen, mindbody. Key to her core paradigms of mindbody and emergence are interaction, mutuality, and codetermination. It is thus altogether fitting that Hayles, drawing on Don Ihde's phenomenological theory of technology, emphasizes three different kinds of interrelationships among the human, technology, and the environment as they are each foregrounded in pieces by Simon Penny, Alan Dunning and Paul Woodrow, and Victoria Vesna—movement, perception, and the cultural construction of relationality respectively.

All three of these artworks reject the aesthetic of realism. All three resist the fantasies of immortality and the transcendence of the body that are so often attributed to cybertechnologies. But what especially interests me is Hayles's observation that all three art works are examples of distributed cognitive systems. "The complexity of the collaborations between many different humans and many different intelligent machines," she writes, "indicates that in a deep sense all of these projects are distributed cognitive systems." These intelligent machines and these software programs are to be understood, in her wonderful phrase, as our "silicon collaborators" (63).

In *How We Became Posthuman*, Hayles observes that the neurologist Antonio Damasio similarly stresses the error of the concept of the disembodied mind. "The Cartesian idea of a disembodied mind may well have been the source, by the middle of the twentieth century," Damasio suggests in *Descartes' Error: Emotion, Reason, and the Human Brain*, "for the metaphor of mind as software program," a metaphor that is misleading because it does not recognize the importance of embodiment.[1] Even more significant in my view is Damasio's emphasis on the centrality of the emotions and feelings as constitutive of knowledge, or as having what I call a cognitive edge. Critical to his argument is "that feelings are a powerful influence on reason, that the brain systems required by the former are enmeshed in those needed by the latter, and that such specific systems are interwoven with those which regulate the body" (245).

Damasio's work has had a powerful influence on many media artists as well as on cultural critics. Dunning and Woodrow, for instance, quote him approvingly.[2] Thus as an extension or supplement to Hayles's emphasis on systems of distributed or collaborative cognition, I would like to propose that we also consider systems of distributed emotion—and not only as a concept but also as a phenomenology. This particular phenomenology of technology would be expressed specifically in terms of the emotions and moods and not in terms of the sensations of movement, visuality, and aurality that we more often associate with virtual reality artworks and that are indeed central to the pieces Hayles considers.

In *Virtualities: Television, Media Art, and Cyberculture*, Margaret Morse discusses several virtual reality artworks from precisely this perspective. A prime example is Luc Courchesne's *Family Portrait* (1993), a piece that allows a person visiting the installation to "talk" with one of nine family members who are presented as full-length spectral portraits. One can interact with members of this virtual family, becoming more intimate with them on a psychological level. But at the same time, as Morse has described to me in conversation, the piece is disconcerting as well. Because these spectral portraits are programmed to respond to each other as well as to the human presence in their midst, virtually looking through the human visitor, one can feel in a peculiar way rejected, left out of the family circle. This sense of rejection is of course a particularly human response (we might call it Freudian) to the family in general. Thus here the cyberworld and the humanworld are connected by the mutual bonds of the emotions. If technology is often understood as an extension of the human body, as a prosthesis that strengthens the body, here emotions serve as a prosthesis connecting the technological world of virtual reality with the humanworld.[3] Indeed these emotions might be said to be themselves emergent, arising from the interaction of the cyberworld and the humanworld.

Morse concludes *Virtualities* by suggesting that virtual artworks may serve to develop the all-important capacity of empathy, noting that she sees "a growing concern with the nurture of artificial life as a kind of exercise in the practice and refinement of empathy."[4] I would add that the mutual development of empathy between nonhuman cyborgs and humans has long been a theme of a wide range of novels and films. A descendent of the computer, the nonhuman cyborg is often figured as a hybrid organism endowed with feeling, as an artificial entity that becomes an organism (or humanlike) precisely because of its capacity for feeling, which implies the ability to form connections with humans; crucial are the binding emotions of sympathy and love. Equally implied is emotional reciprocity on the part of humans.

Interrelationship and mutuality are thus at stake, as is, at base, empathy.

Steven Spielberg's *AI* is only one of the most recent, if unfortunately banal, examples. We can also point to Arthur C. Clarke's first three novels in his *Space Odyssey*, where for both HAL and Bowman the capacity to respond to a situation with sustained feeling, and not just with logic or reason, is ultimately figured as an evolutionary strength, as a critical component of life, whether it is at base biological, electronic, or spiritual. Another major example is Philip K. Dick's celebrated 1968 novel *Do Androids Dream of Electric Sheep?*, which was made into the now-classic film *Blade Runner* in 1982. *Blade Runner* presents us with a model of emotional life arising out of organic embodiment, with emotional intelligence coming to complement artificial intelligence. Emotions arise in the replicants not only by virtue of the development of intersubjective ties but also spontaneously, as it were, by virtue of their embodiment. *Silent Running*, a 1977 cult science fiction film directed by Douglas Trumbell and starring Bruce Dern, and *Solo*, a 1996 action film directed by Norberto Barba, are also cases in point.

In all of these texts, what is thematized is the process of technocultural feedback loops generating emotional growth. The emergence of intersubjectivity between the humanworld and the technological world (represented by replicants and nonhuman cyborgs) results in a form of intelligence—emotional intelligence—that is not only resourceful in a multitude of ways but is also deeply benevolent (these texts are profoundly utopian in spirit). What is ultimately represented, then, is a system of distributed emotional intelligence where the human mindbody has profoundly meaningful ties of feeling to the cyberworld, feelings that are reciprocated.

1   Antonio R. Damasio, *Descartes' Error: Emotion, Reason, and the Human Brain* (New York: G. P. Putnam, 1994), p. 250.

2   "The real-time rendering engines," Dunning and Woodrow write in "Einstein's Brain," "provide a space in which the spontaneous processes of being in the world are made evident, generating what neuroscientist Antonio Damasio describes as a 'dispositional representation of the self that is in the process of changing as the organism responds to an object.' " Alan Dunning and Paul Woodrow, "Einstein's Brain,"http://www.ucalgary.ca/~einbrain/EBessay.htm, footnote 6.

3   See Kathleen Woodward, "Prosthetic Emotions," in *Emotion in Postmodernism*, ed. Gerhard Hoffmann and Alfred Hornung (Heidelberg: C. Winter, 1997), pp. 95–108.

4   Margaret Morse, *Virtualities: Television, Media Art, and Cyberculture* (Bloomington: Indiana University Press, 1998), p. 210.